AF614712

NEUROMETHODS

Series Editor
Wolfgang Walz
University of Saskatchewan
Saskatoon, Canada

For further volumes:
http://www.springer.com/series/7657

MicroRNA Technologies

Edited by

Min Jeong Kye

University of Cologne
Institute of Human Genetics
Cologne, Germany

Editor
Min Jeong Kye
University of Cologne
Institute of Human Genetics
Cologne, Germany

ISSN 0893-2336 ISSN 1940-6045 (electronic)
Neuromethods
ISBN 978-1-4939-7173-2 ISBN 978-1-4939-7175-6 (eBook)
DOI 10.1007/978-1-4939-7175-6

Library of Congress Control Number: 2017942733

Printed on acid-free paper

This Humana Press imprint is published by Springer Nature
The registered company is Springer Science+Business Media LLC
The registered company address is: 233 Spring Street, New York, NY 10013, U.S.A.

Preface to the Series

Experimental life sciences have two basic foundations: concepts and tools. The *Neuromethods* series focuses on the tools and techniques unique to the investigation of the nervous system and excitable cells. It will not, however, shortchange the concept side of things as care has been taken to integrate these tools within the context of the concepts and questions under investigation. In this way, the series is unique in that it not only collects protocols but also includes theoretical background information and critiques which led to the methods and their development. Thus it gives the reader a better understanding of the origin of the techniques and their potential future development. The *Neuromethods* publishing program strikes a balance between recent and exciting developments like those concerning new animal models of disease, imaging, in vivo methods, and more established techniques, including, for example, immunocytochemistry and electrophysiological technologies. New trainees in neurosciences still need a sound footing in these older methods in order to apply a critical approach to their results.

Under the guidance of its founders, Alan Boulton and Glen Baker, the *Neuromethods* series has been a success since its first volume published through Humana Press in 1985. The series continues to flourish through many changes over the years. It is now published under the umbrella of Springer Protocols. While methods involving brain research have changed a lot since the series started, the publishing environment and technology have changed even more radically. Neuromethods has the distinct layout and style of the Springer Protocols program, designed specifically for readability and ease of reference in a laboratory setting.

The careful application of methods is potentially the most important step in the process of scientific inquiry. In the past, new methodologies led the way in developing new disciplines in the biological and medical sciences. For example, Physiology emerged out of Anatomy in the nineteenth century by harnessing new methods based on the newly discovered phenomenon of electricity. Nowadays, the relationships between disciplines and methods are more complex. Methods are now widely shared between disciplines and research areas. New developments in electronic publishing make it possible for scientists that encounter new methods to quickly find sources of information electronically. The design of individual volumes and chapters in this series takes this new access technology into account. Springer Protocols makes it possible to download single protocols separately. In addition, Springer makes its print-on-demand technology available globally. A print copy can therefore be acquired quickly and for a competitive price anywhere in the world.

Saskatoon, Canada ***Wolfgang Walz***

Preface

Impeccable regulation of gene expression is crucial for cellular differentiation and maintaining their identities. To obtain faultless gene expression system, eukaryotic cells developed numerous mechanisms to regulate their gene expression tightly. Among them, microRNA was proposed as a molecule for fine-tuning protein synthesis. Since the function of microRNA was first reported in *C. elegans* model decades ago, it has been intensively studied in various model systems. Functional studies indicate that microRNAs are important for a broad range of cellular and developmental processes such as cell cycle, stem cell maintenance, immune reaction, energy metabolism as well as memory formation. Moreover, increasing numbers of reports pile up on the function of microRNAs. However, it seems that progress of microRNA research in neuroscience is relatively slower than in other fields. This can be due to the complexity of the nervous system as well as the technical difficulties of studying cells in the nervous system.

In this book, we introduce different approaches to understand the function of microRNA and mechanisms regulating microRNA expression in the nervous system. In addition, we also present newly established techniques to study the role of microRNAs in the nervous system. We would be pleased if this book can help students and researchers, who are curious about microRNAs in the nervous system.

Cologne, Germany *Min Jeong Kye*

Contents

Contributors

XIAOWEN BAI • *Departments of Anesthesiology and Physiology, Medical College of Wisconsin, Milwaukee, WI, USA*

MARK A. BEHLKE • *Integrated DNA technologies, Integrated DNA Technologies, Inc., Coralville, IA, USA*

ZELJKO J. BOSNJAK • *Department of Anesthesiology, Medical College of Wisconsin, Milwaukee, WI, USA; Department of Physiology, Medical College of Wisconsin, Milwaukee, WI, USA*

FEDERICO DAJAS-BAILADOR • *School of Life Sciences, University of Nottingham, Nottingham, UK*

ALEXANDER DEITERS • *Department of Chemistry, University of Pittsburgh, Pittsburgh, PA, USA*

ANDRII DOMANSKYI • *Institute of Biotechnology, University of Helsinki, Helsinki, Finland*

ANNA EMDE • *Department of Molecular Genetics, Weizmann Institute of Science, Rehovot, Israel*

PATRICIA P. GARCEZ • *The Francis Crick Institute-Mill Hill Laboratory, London, UK; Instituto de Ciências Biomédicas, Universidade Federal do Rio de Janeiro, Rio de Janeiro, Brazil*

FRANCOIS GUILLEMOT • *Neural Stem Cell Biology Laboratory, The Francis Crick Institute-Mill Hill Laboratory, London, UK*

CAROLINE HICKS • *ReNeuron, Pencoed Business Park, Bridgend, UK*

ERAN HORNSTEIN • *Department of Molecular Genetics, Weizmann Institute of Science, Rehovot, Israel*

GEUN-WOO JIN • *Bioengineering Department, College of Engineering, Temple University, Philadelphia, PA, USA*

VLADIMIR JOVASEVIC • *Department of Psychiatry and Behavioral Sciences, Northwestern University, Chicago, IL, USA*

HAK HEE KIM • *Nemours Biomedical Research, Alfred I. duPont Hospital for Children, Wilmington, DE, USA*

WITOLD KONOPKA • *Molecular Biology of the Cell I, German Cancer Research Center (DKFZ), Heidelberg, Germany; Laboratory of Animal Models, Neurobiology Center, Nencki Institute of Experimental Biology of the Polish Academy of Sciences, Warsaw, Poland*

MIN JEONG KYE • *Institute of Human Genetics, University of Cologne, Cologne, Germany*

MINGYU LIANG • *Department of Physiology, Medical College of Wisconsin, Milwaukee, WI, USA*

WEILI MA • *Bioengineering Department, College of Engineering, Temple University, Philadelphia, PA, USA*

LAURENT NGUYEN • *Interdisciplinary Cluster for Applied Genoproteomics (GIGA-R), University of Liège, Liège, Belgium; GIGA-Neurosciences, University of Liège, Liège, Belgium*

JESSICA M. OLSON • *Department of Anesthesiology, Medical College of Wisconsin, Milwaukee, WI, USA; Department of Physiology, Medical College of Wisconsin, Milwaukee, WI, USA*

MONICHAN PHAY • *Nemours Biomedical Research, Alfred I. duPont Hospital for Children, Wilmington, DE, USA; Department of Biological Sciences, University of Delaware, Newark, DE, USA*

PIERRE-PAUL PRÉVOT • *Interdisciplinary Cluster for Applied Genoproteomics (GIGA-R), University of Liège, Liège, Belgium; GIGA-Neurosciences, University of Liège, Liège, Belgium*

JELENA RADULOVIC • *Department of Psychiatry and Behavioral Sciences, Northwestern University, Chicago, IL, USA*

IRIT REICHENSTEIN • *Department of Molecular Genetics, Weizmann Institute of Science, Rehovot, Israel*

NATALIA RIVKIN • *Department of Molecular Genetics, Weizmann Institute of Science, Rehovot, Israel*

JOHN D. SINDEN • *ReNeuron, Pencoed Business Park, Bridgend, UK*

LARA STEVANATO • *ReNeuron, Pencoed Business Park, Bridgend, UK*

ESTHER T. STOECKLI • *Institute of Molecular Life Sciences and Neuroscience Center Zurich, Zurich, Switzerland*

WON H. SUH • *Bioengineering Department, College of Engineering, Temple University, Philadelphia, PA, USA*

LAVANIYA THANABALASUNDARAM • *ReNeuron, Pencoed Business Park, Bridgend, UK*

DANIELLE TWAROSKI • *Department of Anesthesiology, Medical College of Wisconsin, Milwaukee, WI, USA; Department of Physiology, Medical College of Wisconsin, Milwaukee, WI, USA*

ILYA A. VINNIKOV • *School of Life Sciences and Biotechnology, Shanghai Jiao Tong University, Shanghai, People's Republic of China*

MARIE-LAURE VOLVERT • *Interdisciplinary Cluster for Applied Genoproteomics (GIGA-R), University of Liège, Liège, Belgium; GIGA-Neurosciences, University of Liège, Liège, Belgium*

NICOLE H. WILSON • *Queensland Brain Institute, University of Queensland, Brisbane, QLD, Australia*

YASHENG YAN • *Departments of Anesthesiology and Physiology, Medical College of Wisconsin, Milwaukee, WI, USA*

TAL YARDENI • *Department of Molecular Genetics, Weizmann Institute of Science, Rehovot, Israel*

SOONMOON YOO • *Nemours Biomedical Research, Alfred I. duPont Hospital for Children, Wilmington, DE, USA; Department of Biological Sciences, University of Delaware, Newark, DE, USA*

Neuromethods (2017) 128: 1–10
DOI 10.1007/7657_2016_6

Published online: 15 November 2016

Protocol for miRNA In Situ Hybridization on Mouse Spinal Cord

Irit Reichenstein and Eran Hornstein

Abstract

microRNAs (miRNAs) are key posttranscriptional regulators of gene expression and are involved in a variety of processes in the central nervous system. miRNAs play numerous roles in neuro-development and are crucial for the proper homeostatic function of neurons in adults. The detection of miRNAs in specific cell types sheds light on differential regulation of their expression and on the potential functional roles of miRNAs in the brain. In this chapter, we present a simple and robust protocol for the detection of miRNAs by in situ hybridization in formalin-fixed paraffin-embedded spinal cords of mature mice and developing embryos.

Keywords: In situ hybridization, Spinal cord, microRNA, miRNAs

1 Introduction

miRNAs are abundant within the central nervous system (CNS) and play diverse roles in neuronal development, including neural stem cell differentiation, synapse formation, and dendritogenesis ([1, 2]). In the adult brain, miRNAs contribute to the maintenance of neuron cell identity and govern excitability [3], plasticity [4], and neuron-to-astrocyte communication [5].

The expression of individual miRNAs may be restricted to specific domains of the CNS, to certain cell types, or to subcellular compartments. For example, miR-134 is enriched in dendrites [6], whereas miR-124 is enriched in the cell soma [7]. The detection of miRNAs in specific cell types may provide insight into the function of these miRNAs in healthy neurons and to the consequences of their dysregulation in disease conditions. Recently, there has been a growing interest in the involvement of miRNA dysregulation in motor neuron pathologies, such as spinal muscular atrophy (SMA; [8]), amyotrophic lateral sclerosis (ALS; [9–12]), and spinal cord injury (reviewed in [13]).

Here, a simple and robust protocol for miRNA in situ hybridization on formalin-fixed paraffin-embedded spinal cords of mature mice and developing embryos is described. The protocol is based on previously reported protocols [14–16] and is intended for analysis of miRNAs specifically within the spinal cord.

2 Methods

To minimize RNase contamination, all solutions are prepared with DEPC-treated water. Surfaces are cleaned with RNase Zap (Life technologies). Glassware and staining racks are cleaned by soaking overnight in 0.1 % SDS, and then for 30 min in 0.1 M NaOH, and finally by washing with DEPC-treated water. Unless otherwise indicated, all procedures are performed at room temperature in a chemical hood, wearing gloves and lab coat. Recipes for all solutions used in this protocol are listed in Table 1.

2.1 Spinal Cord Preparation

2.1.1 Transcardial Perfusion

This step enables rapid tissue fixation since the fixative is injected into the heart and cardiac output enables fixative spreading through the entire body. Furthermore, this step washes away erythrocytes from blood vessels. Adult mice are deeply anesthetized with ketamine/xylazine (0.25 ml, 10 % (vol/vol), administered i.p.). Transcardial perfusion is performed as in [17] with 10 ml of PBS, followed by 100 ml of 2.5 % paraformaldehyde (PFA).

2.1.2 Fixation and Preparation of Paraffin Blocks

Following transcardial perfusion, spinal cords are removed, along with the vertebrae and rib cage, and incubated for 48 h in 2.5 % PFA at 4 °C with mild agitation. The spinal cords are then dissected out of the vertebrae and cut at L4–L5 level and incubated in 1.25 % PFA for an additional 24 h at 4 °C, dehydrated in graded ethanol: PBS solutions (30 %; 50 %, 80 %, 90 %, 100 % at steps of 30 min), and incubated in 100 % ethanol two additional times for 30 min, and twice in xylene for 30 min each. The tissue is then incubated with liquid paraffin for at least 3 h and embedded into a paraffin block.

To create transverse sections, blocks are trimmed, rotated 90°, re-embedded, and kept at −20 °C until sectioning.

2.1.3 Embryonic Mouse Spinal Cords

Pregnant females are euthanized by CO_2 on postcoital day 13.5, counting the morning on which vaginal mucus plug is identified as day 0.5. The uterus is surgically removed and placed in a Petri dish with cold PBS. Embryos are further dissected from amniotic sacs, submerged in a Petri dish with cold PBS, and incubated for 24 h in 4 % PFA at 4 °C. Embryos are then dehydrated in graded ethanols, washed, and processed into paraffin blocks as previously described (Section 2.1.2).

2.2 Tissue Sectioning

5–7 μM sections are collected and transferred to a DEPC-treated water bath and heated to 45 °C in order to allow tissue relaxation. Tissue sections are mounted on Superfrost Plus slides (Menzel-Glaser). Slides are dried overnight on a heating block at 37 °C and either used immediately or kept at 4 °C for up to 2 weeks.

Table 1
Solutions for in situ hybridization

Reagent	Volume	Reagent setup
DEPC-treated water		
Diethyl pyrocarbonate (DEPC) (Sigma Aldrich cat# D5758)	1 ml	Stir overnight and autoclave
Double-distilled water	1 L	
DEPC-treated PBS		
Diethyl pyrocarbonate (DEPC) (Sigma Aldrich cat# D5758)	1 ml	Stir overnight and autoclave
PBS 10× (Biological Industries Cat#02-023-5A)	100 ml	
Double-distilled water	900 ml	
Proteinase K buffer 10×		
DEPC-treated Tris–HCl 1 M pH 7.4	500 μl	–
DEPC-treated EDTA 0.5 M	200 μl	
DEPC-treated NaCl 5 M	20 μl	
DEPC-treated water	9.28 ml	
1-Methylimidazole solution		
1-Methylimidazole (Sigma Aldrich cat# 336092)	5 ml	Adjust to pH = 8 using HCl and complete the volume to 500 ml using DEPC-treated distilled water
DEPC NaCl 5 M	30 ml	
DEPC-treated water	400 ml	
EDC solution		
N-(3-dimethylaminopropyl)-*N*′-ethylcarbodiimide (EDC) (Sigma Aldrich cat# 39391)	127 μl	EDC is reactive with oxygen. Aliquot stock solution in an oxygen-free environment (Argon hood) and store at −20°C until use
1-Methylimidazole solution	10 ml	
Acetylation solution		
Triethanolamine (Sigma Aldrich Cat# T1377)	3.3 ml	Mix well triethanolamine with DEPC-treated water; only then add acetic anhydride and mix well
Acetic anhydride (Sigma Aldrich cat# A6404)	625 μl	
DEPC-treated water	246 ml	
1× Hybridization buffer		
2× Hybridization buffer (Exiqon cat# 20822)	1 ml	–
DEPC-treated water	1 ml	
TN buffer		
DEPC-treated Tris 1 M pH 7.5	200 ml	–
DEPC-treated NaCl 5 M	60 ml	
DEPC-treated water	1740 ml	

(continued)

Table 1
(continued)

Reagent	Volume	Reagent setup
TNT buffer		
TN buffer 20 % Tween 20	1 L 2.5 ml	–
TNB buffer		
TN buffer 10 % Blocking reagent (Roche cat# 11096176001)	19 ml 1 ml	–
3 % Hydrogen peroxidase solution		
30 % Hydrogen peroxide (Merck Millipore cat# 822287) TNT	25 ml 225 ml	–
B3 buffer		
DEPC-treated Tris 1 M pH9.5 DEPC-treated NaCl 5 M DEPC-treated $MgCl_2$ 1 M DEPC-treated water	25 ml 5 ml 12.5 ml 207.5 ml	Filter through 0.45 μm filter
Formamide/SSC solution		
Formamide (Merck Millipore cat# 109684) 20× SSC (Life Technologies cat# AM9763) DEPC-treated water	30 ml 15 ml 15 ml	–

2.3 Deparaffinization

From this point on, unless mentioned otherwise, the following steps are performed in staining racks. Spinal cord sections are incubated twice in xylene for 5 min each, and then hydrated through ethanol solutions to PBS in as follows: 10 immersions of the slide staining rack in 100 % ethanol; 5-min incubation in 100 % ethanol; 10 immersions in 96 % ethanol; 5-min incubation in 96 % ethanol; 10 immersions in 70 % ethanol; 5-min incubation in 70 % ethanol; 5-min incubation in PBS. All procedures are performed at room temperature in a chemical hood.

2.4 Post-fixation and Acetylation

Sections are fixed again in 4 % PFA for 10 min at room temperature and washed three times with PBS. Each slide is treated with 300 μl of proteinase K (exiqon), which is diluted in 1× Proteinase K buffer to a final concentration of 2 μg/ml. The slides are then incubated for 8 min inside a humidity chamber. Sections are washed for 5 min in PBS, re-fixed for an additional 5 min in 4 % PFA, and washed three more times with PBS. In order to prevent miRNA release and diffusion [18], the sections are fixed with 1-ethyl-3-(3-dimethylaminopropyl)carbodiimide (EDC) as follows: slides are equilibrated by 10 immersions in freshly prepared 1-methylimidazole solution followed by a 10-min incubation in 1-methylimidazole solution, then transferred to a humidity chamber, and treated with 500 μl EDC solution per slide. Slides are incubated for 1 h, and then covered with parafilm (Bemis NA), to reduce dehydration. Next, EDC is washed by two brief immersions in PBS. To reduce background binding of negatively charged probe to the tissue sections [14], sections are incubated for 10 min in freshly prepared acetylation solution, then washed twice for 5 min in PBS, and air-dried for 20 min.

2.5 DIG-Labeled LNA Probes

DIG-labeled locked nucleic acid (LNA) probes are synthesized and purchased from Exiqon. Each probe and LNA-RNA hybrid has a specific melting temperature that is defined by the nucleotide content of the oligos. The melting temperature (T_m) of the specific probe used is provided by the manufacturer.

2.6 Hybridization

Double-DIG-labeled probe (Exiqon) for a specific miRNA sequence is diluted in 1× hybridization buffer to 25 nM, final concentration. A hybridization oven is preheated to a temperature that is 20 °C below the T_m of the specific probe used. The LNA probe is denatured for 4 min at 90 °C and then cooled down on ice. Slides are positioned inside a humidity chamber containing formamide/SSC solution. Then, 100 μl of probe mix is applied to each slide, slides are covered with parafilm (Bemis NA), and the humidity chamber is sealed inside a plastic bag and placed in a preheated hybridization oven for ~14 h. Stringent 7-min washes are performed in preheated SSC buffers, and diluted from a stock buffer of 20× SSC (Life Technologies), at the hybridization temperature: once in 5× SSC, twice in 1× SSC, and twice in 0.2× SSC. If several probes are used in parallel, all slides are washed at the lowest hybridization temperature. Slides are then washed in 0.1× SSC buffer at room temperature, transferred to PBS, and washed in TN buffer for 5 min.

At this time point, two alternative methods for the detection of in situ signal can be used, depending on the expression level of the miRNA of interest. For detection of moderately to highly expressed miRNAs, we recommend using *fluorescent* in situ *hybridization*, which is described in Section 2.7, and can be combined with co-immunodetection, enabling the identifying of specific cell markers. For detection of less abundant miRNAs, see *histochemical* in situ *hybridization* in Section 2.8.

2.7 Fluorescent In Situ Hybridization

2.7.1 Blocking and Incubation with Anti-DIG Antibody

To block endogenous peroxidases slides are incubated for 30 min in 3 % H_2O_2 (diluted in TNT buffer), and then washed for 5 min in TNT buffer. From this point all washing steps are performed in staining racks, while all other incubation steps are performed in a humidity chamber, and slides are covered with parafilm (Bemis NA), to prevent evaporation. Blocking against nonspecific binding of the detection antibody is achieved by a 1-h incubation period in TNB buffer (200 μl per slide) at room temperature. Slides are incubated with anti-DIG POD antibody (Roche) diluted 1:500 in TNB buffer overnight at 4 °C.

2.7.2 Co-immunodetection

For co-immunodetection, an additional antibody recognizing the protein of interest can be introduced at this point. We recommend calibrating the optimal antibody concentration by standard immunofluorescent staining prior to combining protein detection with miRNA in situ hybridization. Following the overnight incubation with the primary antibody, slides are washed three times with TNT buffer, for 5 min each, and then incubated with the relevant secondary antibody in TNB buffer for 45 min at room temperature. Unbound secondary antibody is removed by three repeated washes with TNT buffer.

2.7.3 Fluorescent Detection

For miRNA detection, TSA Plus cyanine 3 system (Perkin-Elmer) is used according to the manufacturer's instructions. Slides are washed three times with TNT buffer, 5 min each.

2.7.4 Fluorescent Nuclear Staining

To visualize nuclei, slides are incubated with 4′,6-diamidino-2-phenylindole (DAPI) 0.2 μg/ml solution (Sigma-Aldrich) for 5 min. This is followed with one TNT wash and one PBS wash for 5 min each. Finally, slides are air-dried in the dark and mounted with Shandon immu-mount (Fisher Scientific).

2.8 Histochemical In Situ Hybridization

For detection of less abundant miRNAs, we present an alternative detection method using NBT/BCIP reagent, which enables control of pigment intensity in response to the incubation period. Following aspiration of TN buffer (see Section 2.5), slides are further washed with TNT buffer for 5 min and blocked in TNB buffer (200 μl per slide) for 1 h at room temperature. Next, slides are incubated with alkaline phosphatase-conjugated anti-DIG antibody (Roche), which is diluted 1:1000 in TNB buffer, overnight at 4 °C in a humidity chamber. Slides are then washed three times with TN buffer and once with B3 buffer, 5 min each. The developer solution BCIP/NBT (Sigma Aldrich) is applied and slides are covered with parafilm (Bemis NA) and placed in a humidity chamber for a minimum of 10 min; slides may be kept for up to 4 days, depending on the miRNA expression levels. BCIP/NBT solution is changed daily. The reaction is terminated by a 5-min wash in PBS. Finally, slides are air-dried and mounted with Entellan (Merck Millipore).

3 Results and Discussion

A descriptive flowchart of the protocol depicting key steps is presented in Fig. 1. miRNAs that are highly expressed can be easily visualized using the fluorescent detection method, which also enables the simultaneous detection of proteins of interest. To

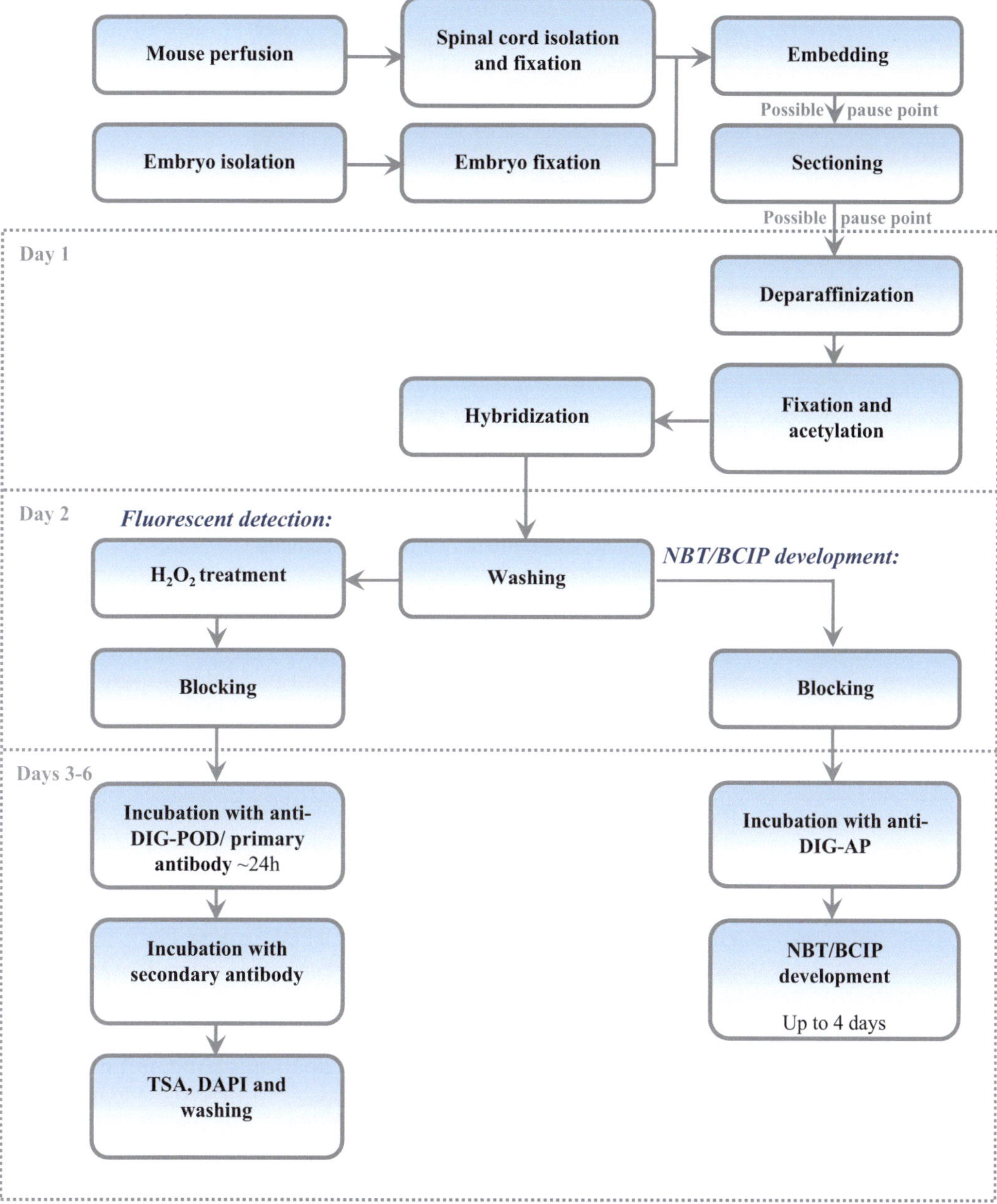

Fig. 1 In situ hybridization flowchart. The flowchart depicts the main steps of the in situ hybridization procedure. Estimated time frame for the different steps is shown. DIG—Digoxigenin; AP—alkaline phosphatase; TSA—tyramide signal amplification

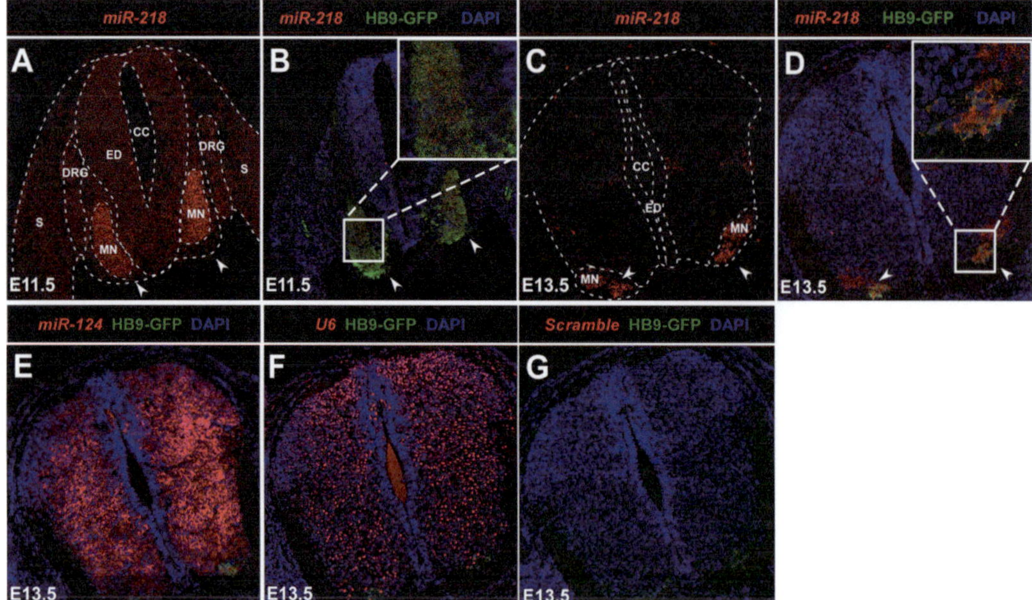

Fig. 2 miRNA fluorescent in situ hybridization on embryonic transverse sections. (**a**–**d**) Immunofluorescent in situ hybridization (FISH) for miR-218 in transverse sections of E11.5 (**a**, **b**) and E13.5 embryos (**c**, **d**) of Hb9-GFP transgenes. miR-218 (*red*) is co-localized with Hb9-GFP (*green*), suggesting motor-neuron-specific expression. (**e**) FISH staining of miR-124, a pan-neuronal miRNA. (**f**) FISH staining of U6 noncoding small RNA depicting nuclear expression pattern. (**g**) FISH staining using a scrambled miRNA probe is used as a negative control. Merged images are counterstained with DAPI depicting nuclei. CC—central canal; ED—ependymal layer; DRG—dorsal root ganglion; MN—motor neurons; S—somites

demonstrate this, we performed a fluorescent in situ hybridization staining of miR-218 on transverse sections of developing embryos (Fig. 2). The staining was performed on a Hb9-GFP transgene mouse embryo; In this mouse line, the motor-neuron-specific Hb9 promoter drives the expression of EGFP, marking all motor neurons in green [19]. The analysis revealed co-localization of miR-218 with Hb9-EGFP, suggesting a motor-neuron-specific expression pattern for miR-218. The expression pattern of miR-218 is consistent with previous reports demonstrating motor-neuron-specific expression of miR-218 in chick [20], mouse [21], and zebrafish [22] spinal cords, as well as with expression of miR-218 hosting genes, Slit2 and Slit3, in E11.5 mouse neural tube [23].

The NBT/BCIP development method enables control of pigment intensity in response to the incubation period and may therefore be advisable with low-abundance miRNAs. Examples for miRNA detection using NBT/BCIP development are presented in Fig. 3.

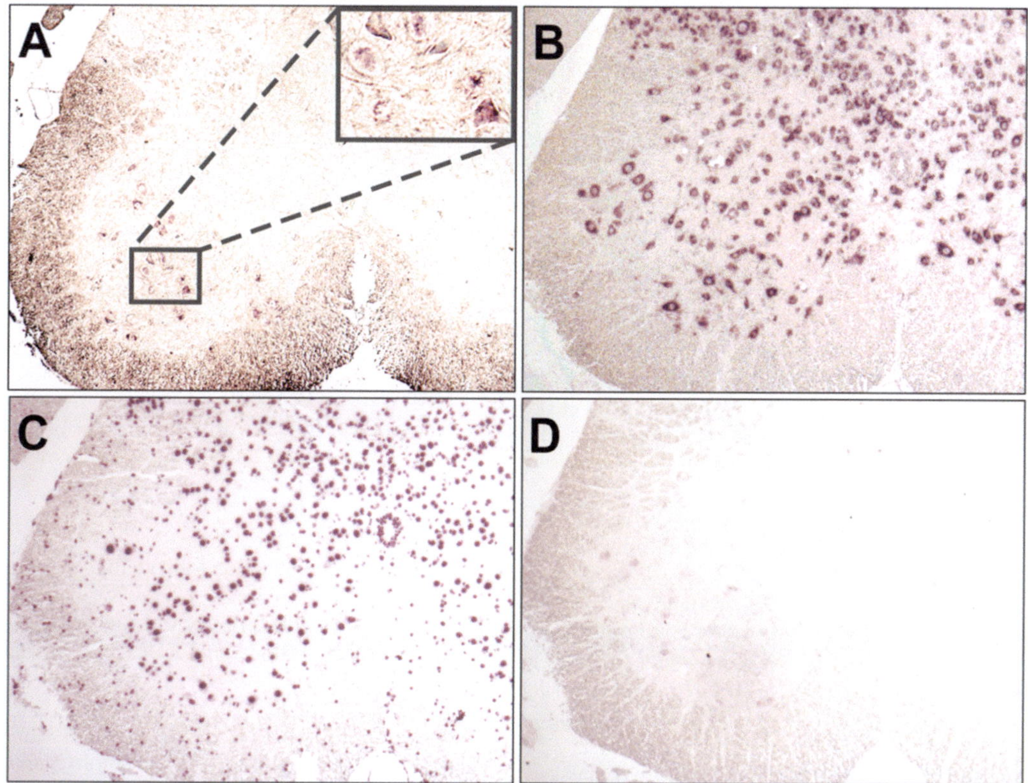

Fig. 3 Non-fluorescent in situ hybridization for miRNAs on mature mouse lumbar spinal cords. (**a**) Staining for the motor-neuron-specific miRNA miR-218. (**b**) Staining for the pan-neuronal miRNA miR-124. (**c**, **d**) U6 snRNA used as a positive control for hybridization. (**c**) and (**d**) scramble probe used as negative control

Acknowledgments

This work was supported by grants to E.H. from the Israel Science Foundation, the ISF Legacy-heritage program, Bruno and Ilse Frick Foundation for Research on ALS, Yeda-Sela, Yeda-CEO fund, Minna-James-Heineman Stiftung through Minerva, ERC consolidator program, Israel Ministry of Trade and Industry "Kamin program," Nella and Leon Benoziyo Center for Neurological Diseases, Y. Leon Benoziyo Institute for Molecular Medicine, and the ALS-Therapy Alliance. Additional funding comes from the Kekst Family Institute for Medical Genetics, David and Fela Shapell Family Center for Genetic Disorders Research, Crown Human Genome Center, Nathan, Shirley, Philip and Charlene Vener New Scientist Fund, Julius and Ray Charlestein Foundation, Fraida Foundation, Wolfson Family Charitable Trust, Adelis Foundation, MERCK UK, Maria Halphen, Estates of Fannie Sherr, Lola Asseof, Lilly Fulop, Teva Pharmaceutical Industries Ltd as part of the Israeli National Network of Excellence in Neuroscience (NNE). Hornstein lab is supported by Dr. Sydney Brenner and Friends.

References

1. McNeill E, Van Vactor D (2012) MicroRNAs shape the neuronal landscape. Neuron 75 (3):363–379
2. Meza-Sosa KF, Pedraza-Alva G, Perez-Martinez L (2014) microRNAs: key triggers of neuronal cell fate. Front Cell Neurosci 8:175
3. Tan CL et al (2013) MicroRNA-128 governs neuronal excitability and motor behavior in mice. Science 342(6163):1254–1258
4. Schouten M et al (2013) microRNAs and the regulation of neuronal plasticity under stress conditions. Neuroscience 241:188–205
5. Morel L et al (2013) Neuronal exosomal miRNA-dependent translational regulation of astroglial glutamate transporter GLT1. J Biol Chem 288(10):7105–7116
6. Schratt GM et al (2006) A brain-specific microRNA regulates dendritic spine development. Nature 439(7074):283–289
7. Kye MJ et al (2007) Somatodendritic microRNAs identified by laser capture and multiplex RT-PCR. RNA 13(8):1224–1234
8. Kye MJ et al (2014) SMN regulates axonal local translation via miR-183/mTOR pathway. Hum Mol Genet 23(23):6318–6331
9. Dini Modigliani S et al (2014) An ALS-associated mutation in the FUS 3′-UTR disrupts a microRNA-FUS regulatory circuitry. Nat Commun 5:4335
10. Marcuzzo S et al (2015) Up-regulation of neural and cell cycle-related microRNAs in brain of amyotrophic lateral sclerosis mice at late disease stage. Mol Brain 8(1):5
11. Wakabayashi K et al (2014) Analysis of microRNA from archived formalin-fixed paraffin-embedded specimens of amyotrophic lateral sclerosis. Acta Neuropathol Commun 2(1):173
12. Emde A et al (2015) Dysregulated miRNA biogenesis downstream of cellular stress and ALS-causing mutations: a new mechanism for ALS. EMBO 34:2633–2651
13. Nieto-Diaz M et al (2014) MicroRNA dysregulation in spinal cord injury: causes, consequences and therapeutics. Front Cell Neurosci 8:53
14. Obernosterer G, Martinez J, Alenius M (2007) Locked nucleic acid-based in situ detection of microRNAs in mouse tissue sections. Nat Protoc 2(6):1508–1514
15. Silahtaroglu AN et al (2007) Detection of microRNAs in frozen tissue sections by fluorescence in situ hybridization using locked nucleic acid probes and tyramide signal amplification. Nat Protoc 2(10):2520–2528
16. Nuovo GJ (2008) In situ detection of precursor and mature microRNAs in paraffin embedded, formalin fixed tissues and cell preparations. Methods 44(1):39–46
17. Gage GJ, Kipke DR, Shain W (2012) Whole animal perfusion fixation for rodents. J Vis Exp 65:pii: 3564
18. Pena JT et al (2009) miRNA in situ hybridization in formaldehyde and EDC-fixed tissues. Nat Methods 6(2):139–141
19. Wichterle, H, Peljto M (2008) Differentiation of mouse embryonic stem cells to spinal motor neurons. Curr Protoc Stem Cell Biol Chapter 1: Unit 1H 1 1–1H 1 9
20. Darnell DK et al (2006) MicroRNA expression during chick embryo development. Dev Dyn 235(11):3156–3165
21. Thiebes KP et al (2015) MiR-218 is essential to establish motor neuron fate as a downstream effector of Isl1-Lhx3. Nat Commun 6:7718
22. Punnamoottil B et al (2015) Motor neuron-expressed microRNAs 218 and their enhancers are nested within introns of Slit2/3 genes. Genesis 53(5):321–328
23. Yuan W et al (1999) The mouse SLIT family: secreted ligands for ROBO expressed in patterns that suggest a role in morphogenesis and axon guidance. Dev Biol 212(2):290–306

Neuromethods (2017) 128: 11–19
DOI 10.1007/7657_2016_1

Published online: 15 November 2016

Protocol for High-Content Screening for the Impact of Overexpressed MicroRNAs on Primary Motor Neurons

Tal Yardeni and Eran Hornstein

Abstract

In this chapter we provide a protocol for the design and usage of automated, high-content microscopy screening that enables the investigation of microRNA (miRNA) impact on primary motor neuron. High-content screening (HCS) platforms facilitate superior precision in research and are scalable to study in parallel multiple genetic, molecular, or cellular conditions. miRNAs are critical for neuronal function and for brain integrity and are considered attractive candidate targets for therapy in many neuropathologies. Therefore, HCS platforms provide a novel paradigm for exploring the impact of miRNA expression, applicable for functional pathways discovery in an academic setting, or towards development of therapeutics in the pharma industry.

Keywords: High-content screening, Primary motor neuron, Automated microscopy, Morphometry, MicroRNA, miRNAs

1 Introduction

MicroRNAs (miRNAs) are endogenous, short, noncoding RNAs. miRNAs silence cellular mRNA targets in a sequence-dependent manner as part of a functional RNA-induced silencing complex (RISC), reviewed in [1]. miRNAs play essential roles in development and function of the central and peripheral nervous systems and miRNA dysregulation is implicated in neurodegenerative diseases (e.g., [2–4]).

Over the last decade, high-content screening (HCS) technology has become increasingly utilized in testing morphological parameters in intact biological systems [5]. The advantages of using HCS platforms reside in precise phenotypic quantification and in reducing the variability often related to culture conditions [6]. HCS technology was already utilized in studies of primary cortical neurons [7] or in studies of motor neurons, which were differentiated from induced pluripotent stem cells (iPSCs) of familial ALS patients [8]. Here, we implemented the platform for studies of primary motor neuron morphology and survival. We calibrated HCS and complemented it with a method for effective transfection of synthetic miRNA mimics into motor neurons, to test candidate miRNA impact on axonal morphologies and survival of primary motor neurons.

2 Methods

2.1 Plate Coating

Three hundred and eighty-four multiwell plates (Griener bio-one, cat# 781091) should be coated with 3 μg/ml poly-ornithine (Sigma-Aldrich) in phosphate-buffered saline (PBS) for at least overnight at 4 °C. Poly-ornithine-coated plates can be kept at 4 °C for several weeks. On the day of plating: wash poly-ornithine three times with PBS. Coat plates with 3 μg/ml Laminin (Gibso) in PBS by incubation at 37 °C for 3–5 h in the tissue culture incubator. When cells are ready (Section 2.4 of this protocol), wash laminin three times with PBS. Keep plate surface humid. Automation, using the washer/dispenser II (GNF system, Table 1), is encouraged.

2.2 Spinal Cord Dissection

Euthanize pregnant female mice on postcoital day 13.5, counting the morning, on which vaginal mucus plug was identified as day 0.5. Aseptic abdomen cleanup with 70 % ethanol prior to laparotomy is encouraged. The bicornuate uterus is surgically removed into a Petri dish with cold PBS. Embryos are further dissected from amniotic sacs, submerged in Petri dish with cold PBS, and kept on ice. Spinal cord dissection is performed under binocular in a 10 cm Petri dish filled with cold PBS. For this, an embryo is positioned with the dorsal side up and the trunk is dissected apart from head, limbs, and tail, using micro-scissors (Fig. 1a). While positioning the embryo with one pair of forceps, by grasping the anterior part of the spinal cord (rhombic lips), a second pair of forceps is used for skinning and exposing the developing spinal cord. For spinal cord separation from adjacent somites, dissect with closed forceps at an angle of 30 °, bilaterally (Fig. 1b), all the way from the head to the tail (rostral to caudal). A practical video-based guide with some variations on this method can be found in [9]. Spinal cords are further sliced to ~0.5 mm fragments. Up to six individual spinal cords can be collected in a 2 ml Safelock Microtube (Eppendorf)

Table 1
Robots

Robot	Maker	Functions
Washer/dispenser II	GNF Systems	Plate cells, wash/dispense
Bravo automated liquid-handling platform	Agilent Technologies	Transfect, wash/dispense
EL406 Washer Dispenser	Bio-Tek	Wash/dispense

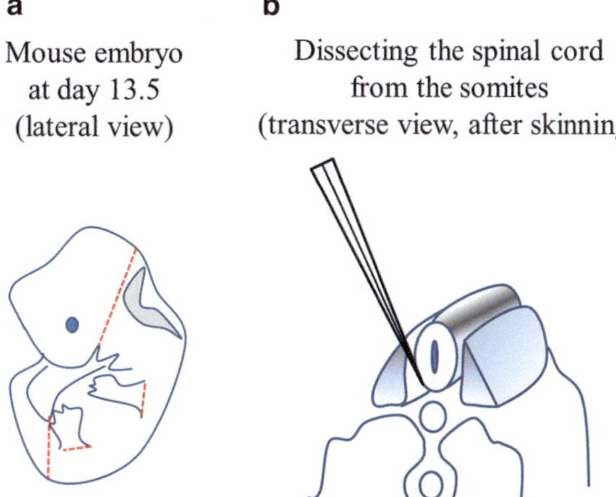

Fig. 1 Spinal cord dissection. An E13.5 mouse embryo is dissected apart from head, limbs, and tail, as demarcated by the *red dashed lines*, using microscissors (**a**). After skinning the embryo and exposing the developing spinal cord, it is separated from adjacent somites, by rostro-caudal movement of closed forceps, at an angle of 30 ° (**b**)

with cold HABG medium (Table 2), which is an artificial neuroprotective cerebrospinal fluid.

2.3 Isolation of Primary Mouse Motor Neurons

This protocol follows Milligan et al. [10], with modifications as described below.

2.3.1 Enzymatic Dissociation

Dissected spinal cord tissue fragments are calibrated to room temperature, and then HABG medium is aspirated in biological hood and replaced with 2 ml of papain (a cysteine hydrolase that is used to isolate cells from the tissue matrix, Table 2). Incubate at 30 °C for 30 min in water bath and gently stir/shake every 2 min. After 30 min, cool microtubes to room temperature.

2.3.2 Mechanic Dissociation

Aspirate Papain and replace with a mix of 1.9 ml HABG with supplementation of 100 ml 10 % BSA and 20 μl DNase at room temperature (Table 2). Gently pipet up and down eight times with a tip of 1 ml. Tip should remain submerged in liquid to avoid foaming. Pellet large tissue debris by gravity and transfer supernatant, which contains dissociated single cells to a clean 15 ml tube. Repeat step for mechanic dissociation twice more with 2 ml HABG (without BSA and DNase).

2.3.3 Density Gradient Separation

OptiPrep™ Density Gradient Medium (Sigma-Aldrich) is used for motor neuron sedimentation. Build discontinuous four-layer OptiPrep gradient according to the recipe in Table 3. Layer the liquids

Table 2
Mediums and solutions

Reagent	Volume	Stock concentration	Vendor	Catalog #
HABG culture medium (50 ml)				
For 12 embryonic spinal cords, divided in 2 tubes				
Hibernate A	48.5 ml	X1	Gibco | Life Technologies	A12475-01
B27	1 ml	X50	Gibco | Life Technologies	12587-010
Penicillin-Streptomycin	0.5 ml	10,000 U/mL	Biological Industries	03-031-1B
L-glutamine	125 μl	200 mM	Biological Industries	03-020-1B
Papain (6 ml)				
For ~18 embryos, 3x 2 ml tubes, containing 6 spinal cords each				
Papain	1 ml	10 mg/ml	Sigma-Aldrich	p4762
Hibernate A (−Ca + 2)	5 ml	X1	BrainBits, LLC	
L-glutamine	15 μl	200 mM	Biological Industries	03-020-1B
HABG BSA DNAse mix (2 ml)				
HABG	1880 μl	1X		
BSA	100 μl	10 %	MP Biomedicals	160069
Dnase	20 μl	1 mg/ml		
Neurobasal medium (50 ml)				
Neurobasal medium	47.5 ml	X1	Gibco | Life Technologies	21103-049
B27	1 ml	X50	Gibco | Life Technologies	12587-010
Horse serum	1 ml	X50	Sigma-Aldrich	H1138
Penicillin-Streptomycin	25 μl	10 mg/ml	Biological Industries	03-031-1B
L-glutamine	125 μl	200 mM	Biological Industries	03-020-1B
GDNF	1 μl	5 μg/ml	PeproTech	450-51
CNTF	5 μl	5 μg/ml	PeproTech	450-50
Blocking solution (100 ml)				
PBS	47.5 ml	X1	Biological Industries	02-020-1A
BSA	2 g		MP Biomedicals	160069
Triton	100 μl	X100	Sigma-Aldrich	X100

of graded densities from layer 1 (bottom) and up to layer 4 (top). Use 1 ml tip and load layers slowly and gently on side of tube to minimize turbulence. 5.5–6 ml of suspension of cells is then gently transferred on top of the OptiPrep gradient. Centrifuge at 800 × g for 15 min at room temperature with no brake. After spinning,

Table 3
OptiPrep gradient

Layer	OptiPrep (μl)	HABG (μl)	Total (ml)
1 (bottom)	173	827	1
2	127	876	1
3	99	901	1
4 (top)	74	926	1

motor neurons are primarily located in the second fraction from bottom (~2–2.5 ml). Aspirate away the upper two phases (~7 ml). Once the top two fractions are removed, the third fraction, that contains motor neurons, is collected in to a clean 15 ml tube. Omit the debris in the lower phase.

2.3.4 Collecting the Isolated Cells

Add 5 ml of fresh HABG to remove remaining OptiPrep and centrifuge at 300 × *g* for 10 min at room temperature. Motor neurons are expected to pellet out at the bottom of the tube. Aspirate HABG/OptiPrep mix and resuspend in 5 ml HABG. Next, inject 1 ml of 4 % BSA in HABG at the bottom of the tube. Centrifuge at 300 × *g* for 10 min at room temperature. Aspirate the supernatant and resuspend cells in 1 ml Neurobasal medium (Table 2). Count cells with hemocytometer and dilute to working concentration of 150 cells/μl in Neurobasal medium.

2.4 Motor Neuron Plating

At this point plates that are pre-coated with poly-ornithine and Laminin should be used (Section 2.1). Use washer/Dispenser II (GNF Systems, Table 1), to plate 7500 cells in 50 μl at each of the 384 multiwells and incubate in 5 % CO_2 at 37 °C. 24 h later, replace 25 μl (50 %) of the medium with fresh Neurobasal medium, using the GNF Systems.

2.5 Automated miRNA Mimics Transfection

miRNA mimics are designed as dsRNA oligonucleotides (Integrated DNA Technologies, Inc. IDT) and encapsulated in Neuro9™ nanoparticles (Precision NanoSystems, Inc. [11]). 24 h after plating remove 25 μl of culture medium from wells by GNF Systems (as in Section 2.4), and add 25 μl mimics in Neuro9 nanoparticles at concentration of 1 ng/μl, reaching final concentration of 0.5 ng/μl in Neurobasal medium using Bravo automated liquid-handing platform (Table 1).

2.6 Automated Staining and Immunocytofluorescence

72 h post-transfection, cells are rinsed with PBS three times using EL406 washer/dispenser (Bio-Tek, Table 1). For fixation, 50 μl of 4 % formaldehyde (ChemCruz) is dispensed onto each well by Bravo (Table 1). Fixation is performed 15 min in room

temperature, followed by three cycles of automated PBS rinsing with the EL406 washer/dispenser. Blocking and permeabilization are performed with 50 μl blocking solution (Table 2) for 20 min at room temperature. Rinse blocking solution 3×PBS EL406 washer/dispenser. 20 μl of primary antibody in PBS with 1 % BSA is added to each well by GNF Systems and incubated for 90 min at room temperature. Antibodies are picked per indication, but a positive reference that demarcates neurons is anti-neuronal class III beta tubulin (Tuj1) antibody (1:1000, MRB-435P, Covance). After incubation with primary antibody, rinse 3× PBS with EL406 washer/dispenser. 20 μl of secondary antibody in PBS with 1 % BSA is added to each well by GNF Systems and incubated for 1 h at room temperature: for example, Cy2-conjugated donkey anti-rabbit IgG (1:200, Jackson ImmunoResearch). 20 μl of 4′6-diamidino-2-phenylindole dihydrochloride (DAPI) (Sigma-Aldrich) at a concentration of 2 μg/ml of PBS with 1 % BSA is added for the last 5 min by GNF Systems. Wash three cycles with PBS. Cover stained cells with 80 μl PBS/well, with EL406 washer/dispenser. At this point, the plate is ready for imaging or can be stored for several days at 4 °C if covered by aluminum foil.

2.7 Automated Microscopic Analysis and Image Processing

For automated fluorescence microscopy ImageXpress Micro equipped with MetaXpress software was used (Molecular Devices). 10× Lens magnification was used with DAPI nuclear stain (filter Ex: 377, Em: 477) and identification of neurons positive to Tuj1 staining was done with FITC-conjugated secondary antibody (filter Ex: 482, Em: 536). In each of the 384 wells on the plate, two sites of 980 μm^2 were automatically defined and captured by the ImageXpress acquisition software. Phenotypic parameters were quantified using Neurite Outgrowth module of MetaXpress (Molecular Devices). Cell bodies were defined as Alexa488-positive objects with dimensions of 80–120 μm width. Cellular outgrowth and processes were defined as Alexa488-positive protrusions of 5 μm width and >10 μm length. Statistical analysis was performed using Student's *t*-test.

3 Results

In this chapter we described the advantages and capability of using HCS to accurately identify miRNAs that regulate motor neuron survival or morphology in an unbiased, high-throughput manner. Primary motor neurons were isolated at mouse embryonic day 13.5. 7500 cells were plated automatically, per microwell, in a 384-multiwell plate. Transfection of miRNA mimics is performed 24 h post-plating (Fig. 2a). miR-142-3p is expressed mainly in hematopoietic cells and hence serves as a miRNA mimic transfection control. To test the transfection system, we used miR-142-3p (miR-142) as a control miRNA and overexpressed it in motor

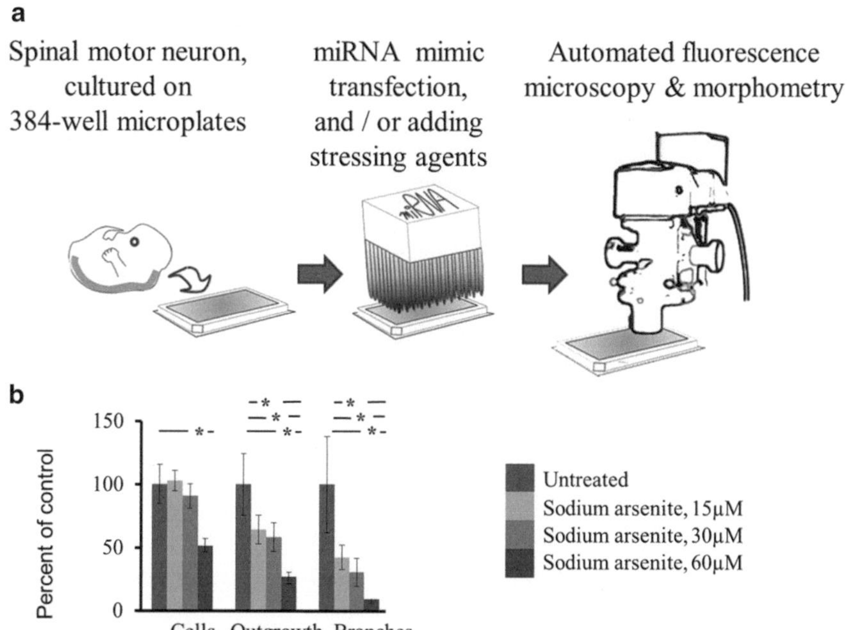

Fig. 2 High-content, automated microscopic analysis of primary motor neurons. Diagram of culturing primary motor neurons from E13.5 mouse embryos on a 384 multiwell plate, followed by miRNA transfection and/or induction of stress. 72 h later, cells were fixed, stained with specific antibodies, and automated fluorescence microscope was performed (**a**). On average, 500 Tuj1+ neurons per field in two fields/well and six wells per treatment were quantified. Cell numbers (Cell), neurite outgrowth per cell (outgrowth), and number of branches per cell (branches) were quantified whether untreated or with serial doses of the stress-inducing compound, sodium arsenite at 15, 30, and 60 μM (**b**)

neurons. The guide sequence of mmu-miR-142-3p is U*G*UAGUGUUUCCUACUUUAUmGmGA.

The passenger strand sequence is 5′C3(spacer)/UmCCmAUmAAmAGmUAmGGmAAmACmACmUAmCA/3′-Cy5.5. "m" refers to a 2′O-methyl RNA and "*" refers to a phosphorothioate internucleotide linkage. A miR-142 mimic or a scrambled control sequence were tested at two concentrations (0.1 and 0.5 ng/μl), and miR-142 expression levels were measured in cell lysates 24, 48, and 72 h post-transfection. At mimic concentration of 0.1 ng/μl, miR-142 was upregulated 30-fold and at 0.5 ng/μl its expression levels were almost 100 times higher than control. This analysis reveals the efficacy of Neuro9 nanoparticles in transfecting dsRNAs into primary motor neurons.

Then sensitivity of the phenotypic assay was calibrated for this platform. To this end, primary motor neurons were treated with sodium arsenite ($NaAsO_2$), a chemical that induces mitochondrial oxidative stress, 24 h post-plating. Sodium arsenite is known to result in neurite outgrowth abnormalities [12, 13] that were characterized by measurement of cell numbers, neurite outgrowth, and number of branches per cell after 72 h (Fig. 2b). Images of the neuron-specific class III beta-tubulin (Tuj1) together with DAPI (for nuclear staining) were captured on an ImageXpress Micro XL wide-field high-content analysis system (Molecular Devices), at 10× magnification, and morphological parameters were quantified with MetaXpress (Molecular Devices) image analysis software. Treatment with increasing $NaAsO_2$ concentrations (15 or 30 μM) yielded a decrease in axon length and in branch numbers. At even higher concentration(60 μM), reduced cell numbers were also observed, suggesting neuronal death (Fig. 2b). Thus, the platform is able to provide accurate automated morphometry.

4 Discussion

In this chapter we presented a method for automated HCS for investigating the impact of miRNAs on primary motor neuron morphologies. HCS platforms have been developed and implicated over the last decade, which provide a superior system for precise and reproducible research [5, 6]. miRNAs are critical for neurons and for brain integrity [14]. Accordingly, dysregulated miRNA expression or function has been reported in neurodegenerative diseases such as Parkinson [2], Alzheimer [15], Huntington [16], and amyotrophic lateral sclerosis (ALS) [3, 17]. Since miRNAs have a key role in the progression and pathogenesis of neurodegeneration, they are considered candidate targets for therapy. HCS platforms such as the one described here, provide a novel paradigm for exploring the impact of miRNA expression on motor neurons in a high-throughput manner and to shed a light on the consequences of manipulation of candidate therapeutic miRNAs ex vivo in relevant neurons.

Acknowledgments

We thank Noga Kozer, Alexander Plotnikov, Haim M. Barr, Shir Katz, Hila Weiss, and Michal Eisenberg (Weizmann Institute of Science) for collaborations on developing the protocols. Raquel Fine and Cherill Banks for editing. This work was supported by grants from the ALS-Therapy Alliance, Israel Science Foundation, Legacy-heritage Fund, Bruno and Ilse Frick Foundation for Research on ALS, Yeda CEO fund, Minna-James-Heineman Stiftung through Minerva, ERC consolidator program, Nella and Leon Benoziyo Center for Neurological Diseases, Y. Leon Benoziyo Institute for Molecular Medicine, the Weizmann Institute of Science and A. Alfred Taubman via IsrALS, Kekst Family Institute for Medical Genetics, David and Fela Shapell Family Center for Genetic Disorders Research, Crown Human Genome Center, Yeda

Sela Center, Helen and Martin Kimmel Institute for Stem Cell Research, the Maurice and Vivienne Wohl Biology Endowment, the Nathan, Shirley, Philip and Charlene Vener New Scientist Fund, Julius and Ray Charlestein Foundation, Celia Benattar Memorial Fund for Juvenile Diabetes, the Wolfson Family Charitable Trust, and Adelis Foundation. TY was supported by postdoctoral fellowship from Teva Pharmaceutical Industries Ltd as part of the Israeli National Network of Excellence in Neuroscience (NNE) and by the postdoctoral fellowship from the Sara Lee Schupf foundation. E.H. is Head of Nella and Leon Benoziyo Center for Neurological Diseases and the lab is further supported by Dr. Sydney Brenner and Friends.

References

1. Bartel DP (2009) MicroRNAs: target recognition and regulatory functions. Cell 136 (2):215–233
2. Kim J et al (2007) A MicroRNA feedback circuit in midbrain dopamine neurons. Science 317(5842):1220–1224
3. Campos-Melo D et al (2013) Altered microRNA expression profile in Amyotrophic Lateral Sclerosis: a role in the regulation of NFL mRNA levels. Mol Brain 6:26
4. Haramati S et al (2010) miRNA malfunction causes spinal motor neuron disease. Proc Natl Acad Sci U S A 107(29):13111–13116
5. Zock JM (2009) Applications of high content screening in life science research. Comb Chem High Throughput Screen 12(9):870–876
6. Gasparri F (2009) An overview of cell phenotypes in HCS: limitations and advantages. Expert Opin Drug Discov 4(6):643–657
7. Kaltenbach LS et al (2010) Composite primary neuronal high-content screening assay for Huntington's disease incorporating non-cell-autonomous interactions. J Biomol Screen 15 (7):806–819
8. Egawa N et al (2012) Drug screening for ALS using patient-specific induced pluripotent stem cells. Sci Transl Med 4(145):145ra104
9. Langlois SD et al (2010) Dissection and culture of commissural neurons from embryonic spinal cord. J Vis Exp. doi:10.3791/1773
10. Milligan C, Gifondorwa D (2011) Isolation and culture of postnatal spinal motoneurons. Methods Mol Biol 793:77–85
11. Rungta RL et al (2013) Lipid nanoparticle delivery of sirna to silence neuronal gene expression in the brain. Mol Ther Nucleic Acids 2:e136
12. Bhagat L et al (2008) Sodium arsenite induces heat shock protein 70 expression and protects against secretagogue-induced trypsinogen and NF-kappaB activation. J Cell Physiol 215 (1):37–46
13. Huang G et al (2013) Death receptor 6 (DR6) antagonist antibody is neuroprotective in the mouse SOD1G93A model of amyotrophic lateral sclerosis. Cell Death Dis 4:e841
14. Nelson PT, Wang WX, Rajeev BW (2008) MicroRNAs (miRNAs) in neurodegenerative diseases. Brain Pathol 18(1):130–138
15. Cogswell JP et al (2008) Identification of miRNA changes in Alzheimer's disease brain and CSF yields putative biomarkers and insights into disease pathways. J Alzheimers Dis 14(1):27–41
16. Sinha M, Mukhopadhyay S, Bhattacharyya NP (2012) Mechanism(s) of alteration of micro RNA expressions in Huntington's disease and their possible contributions to the observed cellular and molecular dysfunctions in the disease. Neuromolecular Med 14(4):221–243
17. Marcuzzo S et al (2015) Up-regulation of neural and cell cycle-related microRNAs in brain of amyotrophic lateral sclerosis mice at late disease stage. Mol Brain 8(1):5

Neuromethods (2017) 128: 21–27
DOI 10.1007/7657_2016_7

Published online: 15 November 2016

Quantification of Dicer Activity in Mammalian Cell Lysates Using a Non-radioactive Fluorescence Method

Anna Emde, Natalia Rivkin, Mark A. Behlke, and Eran Hornstein

Abstract

In this chapter, we provide a protocol for design and usage of an in vitro, cell-free Dicing assay. Our assay is based on previously established methods and improves the ability to quantitatively measure the catalytic efficacy of the Dicer complex under various cellular conditions via non-radioactive, emitted fluorescence signal readout. Because radioactive labeling is substituted by measuring fluorescent signal the method is easy to handle. Additionally, the use of a defined amount of Cy3-quenched double-stranded substrate allows for precise detection of small changes in cellular Dicing complex activity. This experimental approach might be valuable for investigating cellular miRNA biogenesis and activity under various conditions in health and disease.

Keywords: Double stranded RNA, Precursor microRNA, Dicer complex, Enzymatic activity

1 Introduction

miRNAs are approximately 22-nucleotide-long genome-encoded small RNAs that convey posttranscriptional silencing, thereby mediating broad regulatory functions in health and in disease. The processing of mature miRNAs from initial transcripts is mediated by two main complexes: the initial miRNA transcript (pri-miRNA) is subjected to a nuclear processing by the Drosha-Dgcr8 "microprocessor" complex [3], and the resulting intermediate precursor (pre-miRNA) is exported to the cytoplasm, and then further identified and cut by Dicer and its cofactors, yielding a 22-nt mature miRNA [4]. Partners of Dicer are required for efficient pre-miRNA processing and include Argonaute (Ago) [5], protein kinase interferon-inducible double-stranded RNA-dependent activator (Pact), and TAR RNA-binding protein (Trbp) [6]. Dicer and its cofactors load the mature miRNA onto Ago in the RNA-induced silencing complex (RISC), providing sequence-specific silencing activity [7].

miRNAs ensure cellular robustness and homeostasis in development and maintain cellular function and identity also in the adult life [8, 9]. Therefore, dysregulation of miRNA function may contribute to various physiological and pathological conditions such as aging, neurodegeneration, metabolic dysregulation, and cancer by allowing the activation of aberrant pathways that are repressed under normal miRNA activity in younger individuals [10–14].

Several groups established Dicing activity assays, which often contain an immunoprecipitation step of a defined amount of Dicer, the addition of cofactors, and consecutive evaluation of Dicing complex functions [6, 15]. The readout takes place after radioactive labeling and consecutive Northern blotting. These assays are important but limited by the use of radioactive substrate [16, 17]. In addition, some assays are relying on a recombinant Dicer/ recombinant cofactor proteins that are precipitated and washed, potentially eliminating weakly bound cofactors [6, 15].

In the current book chapter, we describe a highly precise assay that combines two advantages:

1. The ability to use full cell lysates, thereby evaluating the actual cellular enzymatic activity of the Dicing complex under a given cellular condition (for example to compare mutant and wild-type conditions, cellular stress conditions, cancer cell lines to their control counterparts, etc.). This approach eliminates a potential bias in the readout by eliminating immunoprecipitation steps, and allowing for the evaluation of the enzymatic activity of intact Dicing complexes (Dicer and its cofactors) in vitro.
2. The second advantage of our assay is the use of a non-radioactive readout system, via a quenched fluorophore, which is released after Dicing of the double strand. This assay provides a large range of linearity with the ability to perform forerunning calibrating steps. In line with the high sensitivity and precision of a fluorophore-based readout, our assay is able to detect even small changes in cellular Dicing activity (10–20 %) with high accuracy.

2 Material and Methods

2.1 Dicing Assay

This assay has been modified on the base on a previously published assay by [1, 2].

Step 1: Annealing protocol for dsRNA (Fig. 7.1)

2.2 Oligo Formulation

GAPDH-1150-S-3′IBFQ 250nmole RNA Oligo(quencher), HPLC purified

GAPDH-1150-S-3′IBFQ:/5Phos/mCmU rCmArU mUrUrC rCrUrG mGrUmA rUmGrA mCrArA rCrGrA mAmU/ 3IABkFQ/

GAPDH-1150-AS-5′Cy3 (fluorescent oligo) 250 nmol RNA Oligo, HPLC purified

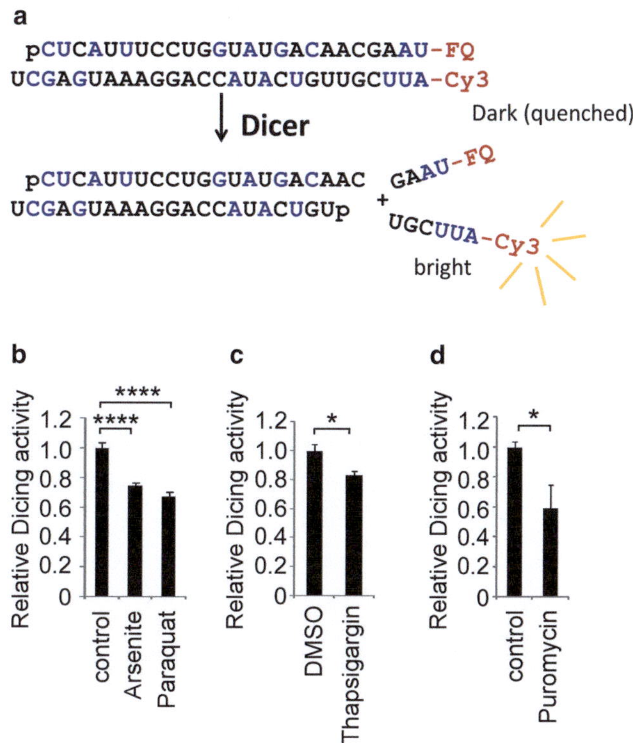

Fig. 7.1 (**a**) Depicted is a schematic flow of the Dicing assay, in which a double-stranded RNA carries a quencher on one strand and a CY3 fluorophore module on the other strand. After dicing of the double-stranded RNA, the quencher releases the CY3 fluorophore module which now can be excitated and emits a signal proportional to the amount of double-stranded RNA diced. *p* = 5′-phosphate; *blue* = 2′OMe RNA; *black* = RNA FQ = Iowa Black FQ (*dark quencher*); Cy3 = fluorescent dye. (**b**) Depicted are results of the relative Dicing complex activity in vitro of NSC-34 cells treated with sodium arsenite (0.5 mM, 60 min), Paraquat (25 μM, 24 h), or control carrier (water). Displayed are results of the relative Dicing complex activity of NSC-34 cells treated with (**c**) control carrier DMSO or Thapsigargin (10 nM, 24 h) and (**d**) control carrier water and Puromycin (1 μg/ml, 24 h). (**b–d**) Depicted are relative Dicing complex activities as estimated by the fluorescent signal emitted by the excitated CY3 fluorophore module after Dicing-activity-dependent release from the quencher module on the complementary siRNA strand. Each group contained at least three biological replicates. Emitted fluorescence signal intensity was measured by the PHERA star FS ™ machine at an excitation/emission frequency of 550 nm/564 nm. Statistics were performed with the two-sided Student's *t*-test including Bonferroni post hoc corrections, comparing control and treatment samples based on the values of the fluorescent intensity signal. Data further normalized to protein levels, quantified by Bradford assay in an aliquot of the sample that was taken for reading fluorescence

GAPDH-1150-AS-5′Cy3:/5Cy3/mAmU mUrCrG rUrUrG mUrCmA rUmArC rCrArG rGrArA rArUmG rAmGmC rU

siRNAs were obtained from Integrated DNA Technologies (IDT) (Syntezza in Israel) as lyophilized RNA oligos.

1. Briefly centrifuge the tubes to bring down the content to the bottom of the tube.

2. Dilute each oligo in RNase-free sterile water (do not use DEPC-treated water) to a final concentration of 500 μM (×10 stock solution).
3. Gently pipette up and down a couple of times to mix well, aliquot into sterile tubes, and store at −20 °C.

2.3 Annealing Double-Stranded RNA Oligos

1. Dilute each siRNA using sterile RNase-free water to a final concentration of 50 μM, from the 10× 500 μM stock solution.
2. Combine 30 μl of the Cy3 ("red") 50 μM, 36 μl of quencher ("black") 50 μM, and 15 μl of annealing buffer (5×) (50 mM Tris pH 7.5, 100 mM NaCl in RNase-free sterile water).
3. Heat the mix to 95 °C for 1 min, then rapidly cool down to 70 °C, and then cool mix down slowly to 37 °C over 3 h.
4. Briefly spin the tube to bring down all droplets from the wall and lid of the tube.
5. Aliquot the annealed dsRNA into RNase-free tubes and store at −20 °C. Do not freeze-thaw more than five times.

2.4 Dicing Assay

1. After exposing cells to the respective treatment conditions, collect cell culture into two 15 ml centrifuge tubes.
2. Centrifuge at 1000× rpm for 5 min at room temperature to pellet the cells, and discard the supernatant.
3. Gently resuspend the cells in 10 ml 1×PBS. Centrifuge at 1000×rpm for 5 min to pellet the cells.
4. Repeat **step 3** twice.
5. Resuspend the cell pellets in 4 ml of ice-cold hypotonic buffer containing 60 mM KCl, and combine both centrifuge tubes into one ice-cold 15 ml centrifuge tube. Centrifuge at 1000×rpm for 5 min at 4 °C, and discard the supernatant.
6. Add 50–80 % volume of ice-cold hypotonic buffer (without KCl). Gently resuspend the cells.
7. Sonicate samples in an ice-cold water bath for 30 s.
8. Transfer the lysate to a 1.5 ml microcentrifuge tube.
9. Centrifuge at 20,000 × g at 4 °C for 30 min.
10. Transfer the supernatant into a new 1.5 ml microcentrifuge tube, and discard the pellets.
11. Assemble the Dicer assay reaction in a volume of 30 μl: 3 μl 10×Buffer K, 3 μl 10 mM ATP, 1.5 μl 0.5 M creatine phosphate, 0.9 μl 1 mg/ml creatine phosphate kinase, 0.6 μl ribonuclease inhibitor (40 units/μl), 3 μl labeled dsRNA (50 nM, IDT), and 18 μl of cell extract.

    It is critical to use freshly dissolved creatine kinase for optimal ATP regeneration.
12. Incubate at 30 °C for 2 h.

13. Add 100 μl of ultrapure water to the 30 μl Dicer assay reaction and transfer each reaction to 10 wells (10 μl each) of 396-well plate. Use a positive control (ssRNA of fluorescent oligo) and a negative control of (dsRNA without cell extract). Read the CY3 fluorescent signal with a fluorometer.

2.5 Stock Solutions

- *Annealing buffer* (5×) (50 mM Tris pH 7.5, 100 mM NaCl in RNase-free sterile water).
- *50× Stock solution of protease inhibitor*: Dissolve one complete protease inhibitor cocktail tablet (Roche) in 1 ml H_2O to make a 50× stock solution. Store aliquots at −20 °C.
- *Hypotonic buffer*: 2 mM $MgCl_2$, 0.1 % β-mercaptoethanol, 10 mM Hepes–KOH, pH 7.0, 1× protease inhibitor. To make 10 ml of the buffer, mix 0.2 ml of 100 mM $MgCl_2$, 0.1 ml of 1.0 M Hepes–KOH, pH 7.0, 10 μl of β-mercaptoethanol, 0.2 ml of 50× stock solution of protease inhibitor, and 9.5 ml of H_2O. *Should be prepared fresh every time.*
- *10× Buffer K*: 200 mM Hepes–KOH, pH 7.0, 20 mM dithiothreitol (DTT), 20 mM $MgCl_2$. Make 10 ml of the 10× buffer by mixing 2 ml of 1 M Hepes, pH 7.0, 0.2 ml of 1 M DTT, 0.2 ml of 1 M $MgCl_2$, and 7.6 ml of H_2O together. *Prepare fresh every time.*
- *0.5 M Creatine phosphate* (disodium salt, Sigma): Dissolve creatine phosphate in ultrapure water, and store aliquots at −20 °C.
- *Storage buffer:* 50 mM Tris–HCl, pH 7.0, 1 mM DTT, 1 μM ATP, 0.1 mM EDTA, and 50 % glycerol.
- *1 mg/ml Creatine phosphate kinase* (Sigma): Dissolve creatine phosphate kinase in storage buffer, and save at −20 °C.
- *RNasin ribonuclease inhibitor*, 40 U/μl (Ambion).
- *TE buffer*: 10 mM Tris–HCl, pH 8.0, 1 mM EDTA. 9. Sample buffer: Deionized formamide/TE (1:1). *Prepare fresh every time.*

3 Results

We were interested in evaluating differential Dicer complex enzymatic activity under conditions of cellular stress. To this end, we applied different cellular stressors to the mixed mouse motor neuron cell line NSC-34: Sodium arsenite (0.5 mM, 60 min), Paraquat (25 μM, 24 h), Thapsigargin (10 nM, 24 h), and Puromycin (1 μg/ml, 24 h) and observed a reduction in Dicing complex activity.

4 Discussion

In this chapter we presented a method for the evaluation for cellular Dicing activity via a fluorescence reporter-based assay. The advantages of this method lay in its non-radioactive approach on one hand, and in the high precision and accuracy on the other. Examination of cellular Dicing activity is crucial, as more and more aging-related diseases such as neurodegeneration, cancer, and metabolic alterations present with aberrant miRNA processing activity and expression [10–14]. Therefore, it is crucial to extend our knowledge of Dicing complex activity alterations under a given cellular condition. Our assay will help researchers to interrogate various questions in this context with high accuracy and precision.

Acknowledgements

We thank Natalia Rivkin, WIS, for excellent assistance and performance in installing the Dicing assay in the Hornstein lab. This work was supported by grants to E.H. from the Israel Science Foundation, the ISF Legacy-heritage program, Bruno and Ilse Frick Foundation for Research on ALS, Yeda-Sela, Yeda-CEO fund, Minna-James-Heineman Stiftung through Minerva, ERC consolidator program, Israel Ministry of Trade and Industry “Kamin program,” Nella and Leon Benoziyo Center for Neurological Diseases, Y. Leon Benoziyo Institute for Molecular Medicine, and the ALS-Therapy Alliance. Additional funding comes from the Kekst Family Institute for Medical Genetics, David and Fela Shapell Family Center for Genetic Disorders Research, Crown Human Genome Center, Nathan, Shirley, Philip and Charlene Vener New Scientist Fund, Julius and Ray Charlestein Foundation, Fraida Foundation, Wolfson Family Charitable Trust, Adelis Foundation, MERCK UK, Maria Halphen, Estates of Fannie Sherr, Lola Asseof, Lilly Fulop, Teva Pharmaceutical Industries Ltd as part of the Israeli National Network of Excellence in Neuroscience (NNE). A.E. was supported by Deutsche Forschungsgemeinschaft. E.H. is Head of Nella and Leon Benoziyo Center for Neurological Diseases and the lab is further supported by Dr. Sydney Brenner and Friends.

References

1. Melo CA, Melo SA (2014) MicroRNA biogenesis: dicing assay. Methods Mol Biol 1182:219–226
2. Collingwood MA et al (2008) Chemical modification patterns compatible with high potency dicer-substrate small interfering RNAs. Oligonucleotides 18(2):187–200
3. Gregory RI et al (2004) The Microprocessor complex mediates the genesis of microRNAs. Nature 432(7014):235–240
4. Bernstein E et al (2001) Role for a bidentate ribonuclease in the initiation step of RNA interference. Nature 409 (6818):363–366

5. Diederichs S, Haber DA (2007) Dual role for argonautes in microRNA processing and post-transcriptional regulation of microRNA expression. Cell 131(6):1097–1108
6. Chendrimada TP et al (2005) TRBP recruits the Dicer complex to Ago2 for microRNA processing and gene silencing. Nature 436 (7051):740–744
7. Gregory RI et al (2005) Human RISC couples microRNA biogenesis and posttranscriptional gene silencing. Cell 123(4):631–640
8. Kim J et al (2007) A MicroRNA feedback circuit in midbrain dopamine neurons. Science 317(5842):1220–1224
9. Hornstein E, Shomron N (2006) Canalization of development by microRNAs. Nat Genet 38 (Suppl):S20–S24
10. Jordan SD et al (2011) Obesity-induced overexpression of miRNA-143 inhibits insulin-stimulated AKT activation and impairs glucose metabolism. Nat Cell Biol 13 (4):434–446
11. Ibanez-Ventoso C et al (2006) Modulated microRNA expression during adult lifespan in Caenorhabditis elegans. Aging Cell 5 (3):235–246
12. Kornfeld JW et al (2013) Obesity-induced overexpression of miR-802 impairs glucose metabolism through silencing of Hnf1b. Nature 494(7435):111–5
13. Liu N et al (2012) The microRNA miR-34 modulates ageing and neurodegeneration in Drosophila. Nature 482(7386):519–23
14. Mori MA et al (2012) Role of microRNA processing in adipose tissue in stress defense and longevity. Cell Metab 16(3):336–347
15. Wilson RC et al (2015) Dicer-TRBP complex formation ensures accurate mammalian microRNA biogenesis. Mol Cell 57 (3):397–407
16. Hutvagner G et al (2001) A cellular function for the RNA-interference enzyme Dicer in the maturation of the let-7 small temporal RNA. Science 293(5531):834–838
17. Lee HY et al (2013) Differential roles of human Dicer-binding proteins TRBP and PACT in small RNA processing. Nucleic Acids Res 41 (13):6568–6576

Neuromethods (2017) 128: 29–42
DOI 10.1007/7657_2016_2

Published online: 31 July 2016

Analysis of MicroRNAs and their Potential Targets in Human Embryonic Stem Cell-Derived Neurons Treated with the Anesthetic Propofol

Danielle Twaroski, Yasheng Yan, Jessica M. Olson, Mingyu Liang, Zeljko J. Bosnjak, and Xiaowen Bai

Abstract

Growing evidence demonstrates that prolonged exposure to general anesthetics, including propofol, induces widespread neuroapoptosis followed by long-term memory and learning disabilities in animal models. The underlying mechanisms of anesthetic-induced neurotoxicity are complex and not well understood. In addition, there is no direct clinical evidence of the detrimental effects of anesthetics in human fetuses, infants, or children. Development of an in vitro neurogenesis system of human stem cells opens up avenues of research for advancing our understanding of the issues of anesthetic-induced developmental neurotoxicity using a relevant human model. One avenue for investigating the mechanisms behind this neuroapoptosis is through evaluation of microRNA expression. MicroRNAs are endogenous, small, noncoding RNAs that negatively regulate target gene expression. MicroRNAs have been implicated to play important roles in many different disease processes, including neurological diseases. Our recent publication showed that among 84 microRNAs screened, propofol exposure altered the expression of 20 microRNAs in human embryonic stem cell (hESC)-derived neurons. Specifically, downregulation of microRNA-21 (miR-21) conferred by propofol played functional roles in the propofol-induced neurotoxicity. In this chapter, we outline (1) the protocol of human neuron differentiation from stem cells and (2) the protocols for analyzing microRNA expression (using our miR-21 study as an example) by quantitative reverse transcription-PCR and screening potential targets of miR-21 by Western blot assay following propofol exposure in stem cell-derived human neurons.

Keywords: Stem cells, Neurons, Propofol, Developmental neurotoxicity, MicroRNAs, PCR

1 Introduction

Growing evidence in rodents and nonhuman primates has demonstrated that prolonged exposure of developing animals to general anesthetics (e.g., propofol) could induce widespread neuroapoptosis followed by long-term memory and learning abnormalities [1–5]. The greatest vulnerability of the developing brain to anesthetics occurs at the time of rapid synaptogenesis in the brain which is also known as the brain growth spurt [6, 7]. In mice, brain growth peaks at approximately 7 days after birth [8]. In humans, synaptogenesis starts during the third trimester of pregnancy and continues for up to 2–3 years after birth [9]. Overall problems and barriers of the current research in this field are the following: (1) So far, there is no direct clinical evidence showing any such effect in pediatric populations and considerable controversy remains as to whether these findings from animal studies are relevant to humans [2, 10, 11]; and (2) the underlying mechanisms of anesthetic neuroapoptosis are complex and just beginning to be understood [12–14].

The largest obstacle when studying the neurotoxic effect of anesthetics in pediatric patients is the development of an appropriate human model. Exposing healthy children to anesthetics in order to study these effects, as is done in animal models, would not be ethical. Direct harvest of neurons from human brains is also currently not feasible, nor would it yield enough cells for the studies. To overcome these barriers, we have developed an in vitro-controlled neurogenesis system in which we are able to derive developing neurons from human embryonic stem cells (hESCs).

hESCs are cells derived from the inner cell mass of a human blastocyst [15]. hESCs are inherently pluripotent, meaning they can differentiate into cells from all three germ layers (ectoderm, endoderm, and mesoderm) and can replicate indefinitely. This, along with their high differentiation efficiency, makes them more advantageous than other stem cell types such as adult stem cells. Thus, the use of hESCs allows for mechanistic-based studies of anesthetic-induced developmental neurotoxicity using a human cell line and eliminates potential concerns regarding the relevancy of animal models to humans.

MicroRNAs are small (approximately 22 nucleotide), endogenous, noncoding RNAs involved in posttranscriptional regulation and mRNA silencing. They have been predicted to affect the expression of a majority of mammalian genes [16–18]. The first microRNA was discovered in 1993 by Lee and colleagues in *C. elegans* [19]. These nucleotide sequences are highly conserved and are believed to be critical components in evolution [20]. MicroRNAs are transcribed in the nucleus by RNA polymerase II and the pri-miRNA product is processed by the enzyme Drosha to produce a single-hairpin structure. This forms the pre-miRNA which is exported out into the cytoplasm by Exportin 5. Once in the cytoplasm, the pre-miRNA is further processed by the enzyme Dicer which removes the hairpin loop and forms the mature microRNA. One strand of mature microRNA then incorporates into the microRNA-induced silencing complex (miRISC) where it can act to induce silencing of its target genes [17]. These small RNAs function by perfect or imperfect complementary binding to target mRNA and induce gene silencing through mRNA degradation or translational repression [16, 21]. MicroRNAs act as negative regulators of their target genes and a single microRNA can have multiple targets. So far, over 1000 microRNAs have been identified in humans and have been shown to play key roles in nearly every cellular process from development through apoptosis and as such they have become attractive therapeutic targets [34, 35].

Several miRNAs have been shown to play important roles in brain injury and ethanol-induced neurotoxicity [22, 23]. However, there is very limited information about the role of microRNAs in anesthetic-induced neurotoxicity. Using stem cell-derived human neurons, our lab was the first to show that propofol induces cell death in a human model of developing neurons and alters the microRNA profile in these neurons [24]. Among 84 microRNAs screened, propofol exposure altered the expression of 20 microRNAs. microRNA-21 (miR-21) was of interest to us due to its established role as an anti-apoptotic factor. miR-21 was one of the first microRNAs discovered in humans and its sequence was found to be highly conserved across species [25]. miR-21 has been shown to protect neurons from ischemic injury [23] and overexpression of miR-21 has been shown to decrease apoptosis in a rat model of traumatic brain injury [26]. Exposure of fetal cerebral cortical-derived neuroepithelial cells to ethanol was shown to suppress miR-21 [22].

In our experiments, we found that propofol downregulated miR-21 expression in the hESC-derived neurons and that overexpression of miR-21 attenuated the propofol-induced neuron death. Since our initial report documenting a role for microRNAs in propofol-induced neuronal cell death in 2014 [24], several additional studies have focused on understanding the role of microRNAs (e.g., miR-34c and miR-124) in anesthetic-induced neurotoxicity [27, 28], suggesting that microRNAs may be very important in this phenomenon. This chapter focuses on detailing the protocols necessary to analyze microRNA expression (using our miR-21 study as an example) by quantitative reverse transcription-PCR (qRT-PCR) and screen their potential targets by Western blot assay following anesthetic exposure in stem cell-derived human neurons.

2 Materials and Methods

2.1 Neural Stem Cell Culture

Neural stem cells (NSCs) generated from hESCs (H1 cell line, WiCell Research Institute Inc., Madison, WI) were cultured in plastic, Matrigel-coated 60 mm dishes in NSC expansion media containing DMEM/F12 supplemented with 2 % B27 without vitamin A, 1 % N2 (Invitrogen, Carlsbad, CA), 1 % nonessential amino acids, 20 ng/mL bFGF, and 1 mg/mL heparin [29–31]. The media was changed every other day and the NSCs were passaged enzymatically every 5 days with Accutase (Innovative Cell Technologies, San Diego, CA).

2.2 Neuron Differentiation

To differentiate the NSCs into neurons, NSCs were cultured in 60 mm Matrigel-coated dishes (500,000 cells/dish) for 2 weeks in neuron differentiation media containing neurobasal media (Gibco, Grand Island, NY) supplemented with 2 % B27, 0.1 μM cyclic adenosine monophosphate, 100 ng/mL ascorbic acid (Sigma-Aldrich, St. Louis, MO), 10 ng/mL brain-derived neurotrophic factor, 10 ng/mL glial cell-derived neurotrophic factor, and 10 ng/mL insulin-like growth factor 1 (PeproTech Inc., Rocky Hill, NJ). The media was changed every other day and after 2 weeks of culture, the cells displayed clear neuronal morphology and were used for the studies. NSCs and differentiated neurons were characterized using immunofluorescence staining by analyzing the NSC- and neuron-specific protein expression.

2.3 Immunofluorescence Staining

NSCs and 2-week-old NSC-derived neurons cultured on Matrigel-coated, glass cover slips were fixed for 30 min at room temperature in 4 % paraformaldehyde. Cells were then washed with phosphate-buffered saline (PBS) three times followed by a 15-min incubation in 0.5 % Triton X-100 (Sigma-Aldrich) in PBS. The cells were then washed with PBS and blocked for 20 min at room temperature with 10 % donkey serum. Following the blocking, the cells were incubated with the primary antibodies against nestin (an NSC marker; 1:200 dilution) (Millipore, Billerica, MA), β-tubulin III (a neuron marker; 1:200 dilution), or doublecortin (an immature neuron marker; 1:1000 dilution) (Abcam, Cambridge, MA) for 1 h in a humidified, 37 °C incubator. The cells were washed three times with PBS and incubated for 45 min at 37 °C with Alexa Fluor 488 or 594 donkey anti-mouse or rabbit immunoglobulin G secondary antibodies (1:1000 dilution) (Invitrogen). The cells were washed with PBS and the cell nuclei were stained with Hoechst 33342 (1 mg/mL) (Invitrogen). Finally, the cover slips were mounted onto glass slides and imaged using a laser-scanning confocal microscope (Nikon Eclipse TE2000-U, Nikon Inc., Melville, NY).

2.4 Propofol Exposure

While assessment of brain concentrations of propofol in humans during the induction and maintenance of anesthesia is difficult, studies have estimated that it ranges from 4 to 20 μg/mL [32–35]. Thus, 2-week-old stem cell-derived neurons were treated with 20 μg/mL of research-grade propofol (Sigma-Aldrich) or equal volume of dimethyl sulfoxide (DMSO, Sigma-Aldrich) as the vehicle control in 60 mm culture dishes (500,000 cells/dish) or 12 mm glass cover slips (100,000 cells/cover slip). A stock solution (40 mg/mL) of propofol was prepared in DMSO. The working concentration of 20 μg/mL of propofol was prepared from the stock in neuron differentiation media. Cells were exposed to propofol for 6 h and then lysed for microRNA and protein analysis.

2.5 Total RNA Extraction

Following exposure to propofol or DMSO, total RNA was extracted using the Qiagen RNeasy mini kit (Qiagen, Valencia, CA). About 1×10^6 cells cultured in 60 mm dishes were rinsed with PBS and lysed with 700 μL QIAzol lysis reagent (Qiagen). The cells were scraped using a cell scraper and the lysate was transferred to a sterile Eppendorf microcentrifuge tube. The lysates were briefly vortexed and placed on the benchtop for 5 min. The cells can also be placed in the −80 °C freezer at this step and processed at a later time. If frozen, the sample should be allowed to thaw completely and sit at room temperature for an additional 5 min. Following the 5-min incubation, 140 μL of chloroform was added to each sample and the lysates were vortexed for 15 s. The samples were placed on the benchtop for 2–3 min and during this time three distinct layers formed (top, clear layer was RNA, middle white layer was protein, and bottom pink layer was organic matter and lysis reagent). The lysates were then centrifuged at 12,000 × g for 15 min at 4 °C. The supernatant (upper, clear layer) was transferred to a new Eppendorf tube. RNA was precipitated from the upper phase by adding 525 μL of 100 % ethanol (or 1.5 volumes) to each sample. The samples were mixed thoroughly by pipetting and 500 μL of each sample was placed onto separate RNeasy spin columns (Qiagen) in collection tubes. The samples were centrifuged at 8000 × g for 15 s at room temperature to get the RNA into the membranes of the columns. This step was repeated with the remaining sample and the flow-through was discarded. The samples were rinsed twice with 500 μL buffer RPE and the samples were centrifuged first for 15 s, and then for 2 min to dry out the column at 8000 × g at room temperature, and the flow-through was discarded. The spin columns were placed in new Eppendorf tubes and the RNA was eluted from the column by adding 30 μL of RNase-free water onto the membrane and centrifuging the samples at 8000 × g for 1 min (Qiagen). The quantity and quality of the RNA were assessed using an Epoch nanodrop spectrophotometer (BioTek Instruments Inc., Winooski, VT). Each RNA sample was then diluted to 100 ng/μL in RNase-free water.

2.6 cDNA Preparation

The RNA was reverse transcribed to cDNA using the miScript II RT kit (Qiagen) following the manufacturer's instructions. A mixture containing 1 μg of RNA (10 μL of 100 ng/μL RNA), 2 μL of 10× miScript nucleics mix, 2 μL of reverse transcriptase mix, 4 μL of 5× HiSpec buffer, and RNase-free water (variable depending on volume of RNA, 2 μL water if 10 μL RNA used) (Qiagen) was prepared. The RT reaction mixture had a final volume of 20 μL and was incubated at 37 °C for 1 h and 95 °C for 5 min to stop the reaction. The RT product was diluted in 200 μL of RNase-free water to give a final RNA concentration of 4.5 ng/μL in each sample. miR-21 expression was assessed using qRT-PCR.

2.7 MicroRNA Analysis by Quantitative Reverse Transcription-PCR (qRT-PCR)

A master mix (25 μL/well) containing the template cDNA (4.5 ng/well), universal primer (2.5 μL/well), RNase-free water (1 μL/well), miScript SYBR Green (12.5 μL/well), and miR-21 primer or Rnu-6 primer (a housekeeping gene) assays (2.5 μL/well) (Qiagen) was prepared according to the manufacturer's directions. The qRT-PCR was run using a BioRad iCycler for 15 min at 95 °C followed by 40 cycles of a three-step denaturation (15 s at 94 °C), annealing (30 s at 55 °C), and extension (30 s at 70 °C). Reverse transcriptase controls were run to ensure primer specificity. Melt curve analysis was conducted by raising the temperature in 0.5 °C stepwise increments from 65 to 95 °C every 5 s to induce product denaturation. The resulting single peak at the appropriate temperature confirmed the specificity of the qRT-PCR primers.

Three cDNA samples were each run in triplicate on the same qRT-PCR run. Cycle threshold (Ct) value was recorded and used for determining the expression level of miR-21. The Ct value indicates the number of PCR cycles necessary to increase reporter dye signal sufficiently high to cross a manually determined threshold value. The earlier the signal passes the threshold value (i.e., the lower the Ct value), the higher the expression of that gene in that sample. The Ct values of the mRNAs in each sample were normalized against Rnu-6 by subtracting the Ct of Rnu-6 from the Ct of miR-21 and presented as ΔCt. The fold change of the miR-21 expression in propofol-treated neurons compared with control group was presented using the following formula: the fold change of the individual gene expression was presented by the $2^{-\Delta\Delta Ct}$ formula in which ΔΔCt = ΔCt of propofol group—ΔCt of no-treatment group. To express the experimental group data as a percentage of the control group, all normalized control values were averaged and each individual normalized control value was divided by the control average and multiplied by 100 to set the control group to 100. Each normalized experimental group value was divided by the control average, multiplied by 100, and averaged to determine the percent of control.

2.8 Western Blot Analysis of Potential Target of miR-21

A single miRNA can target many genes and negatively regulate the translation of target genes. Western blot assay was used to screen potential targets of miR-21 in the neurons treated with propofol. Following exposure to propofol or DMSO (control), the cells (about 1×10^6 cells) were rinsed with PBS, lysed, and sonicated in 500 μL RIPA lysis buffer (Cell Signaling, Danvers, MA) containing phosphatase inhibitor cocktail (Roche Diagnostics). Lysates were centrifuged at 10,000 × *g* for 10 min at 4 °C. Pellets were discarded and the total protein concentration of the supernatants was determined using a DC Protein Assay Reagents Package kit (Bio-Rad, Hercules, CA). The samples were boiled for 5 min at 97 °C. 25 μg of protein was loaded per lane for SDS-polyacrylamide gel electrophoresis separation and then transferred to nitrocellulose membrane. Membranes were blocked with blocking buffer (Thermo Fisher Scientific, Waltham, MA) and then incubated overnight at 4 °C with primary antibodies against potential direct targets of miR-21 [36–38]: programmed cell death 4 (PDCD4) (Rockland), phosphatase and tensin homolog (PTEN), Sprouty 2, or housekeeping gene actin (1:1000 dilution) (Cell Signaling). The membranes were incubated with secondary antibodies conjugated to horseradish peroxidase (Cell Signaling) for 1 h at room temperature and labeled proteins were detected with chemiluminescence detection reagent (Cell Signaling) and obtained on X-ray film. Optical densities were quantified using ImageJ software. The expression levels of PDCD4, PTEN, and Sprouty 2 were normalized by actin expression and the data was reported as % control.

2.9 Statistical Analysis

Results were obtained from at least three independent neuronal differentiations. Values were reported as means ± the standard deviation with normal distributions. Statistical analysis was performed using the Student's *t*-test when comparing two groups. All statistical analysis was performed using the SigmaStat 3.5 software (Systat Software, Inc., San Jose, CA). *P*-values < 0.05 were considered significant.

3 Results and Discussion

3.1 Characterization of NSCs and Developing Human Neurons from NSCs

NSCs displayed a triangle-like shape while cultured on Matrigel-coated dishes (Fig. 1a) and extensively proliferated. They were passaged every 5–6 days for over 12 passages. NSCs were confirmed by the expression of the NSC-specific marker nestin (Fig. 1b). Obvious neuron differentiation could be observed in the culture 3 days after NSCs were cultured in neuron differentiation medium. Following 14 days in differentiation media, the cells displayed a characteristic neuronal morphology with small cell bodies and interconnected projections (Fig. 1a). The differentiated neurons expressed the neuron-specific marker β-tubulin III (Fig. 1b). Based upon the immunostaining, the differentiation protocol was 90–95 % efficient in the generation of neurons.

It has been shown that the period in which developing mammalian neurons are the most vulnerable to anesthetic administration is the period of rapid synaptogenesis or the brain growth spurt [39, 40]. We showed previously that 2-week-old NSC-derived neurons expressed the pre- and post-synaptic markers synapsin I and Homer I, respectively. Punctate synapsin 1 and Homer I signals could be observed around cell bodies and along radially oriented axons [29, 41]. Additionally, these cells also displayed synapse-like structures [31] upon electron microscopic imaging, suggesting that these neurons generated from NSCs were similar to human neurons at morphological and structural levels. The 2-week-old

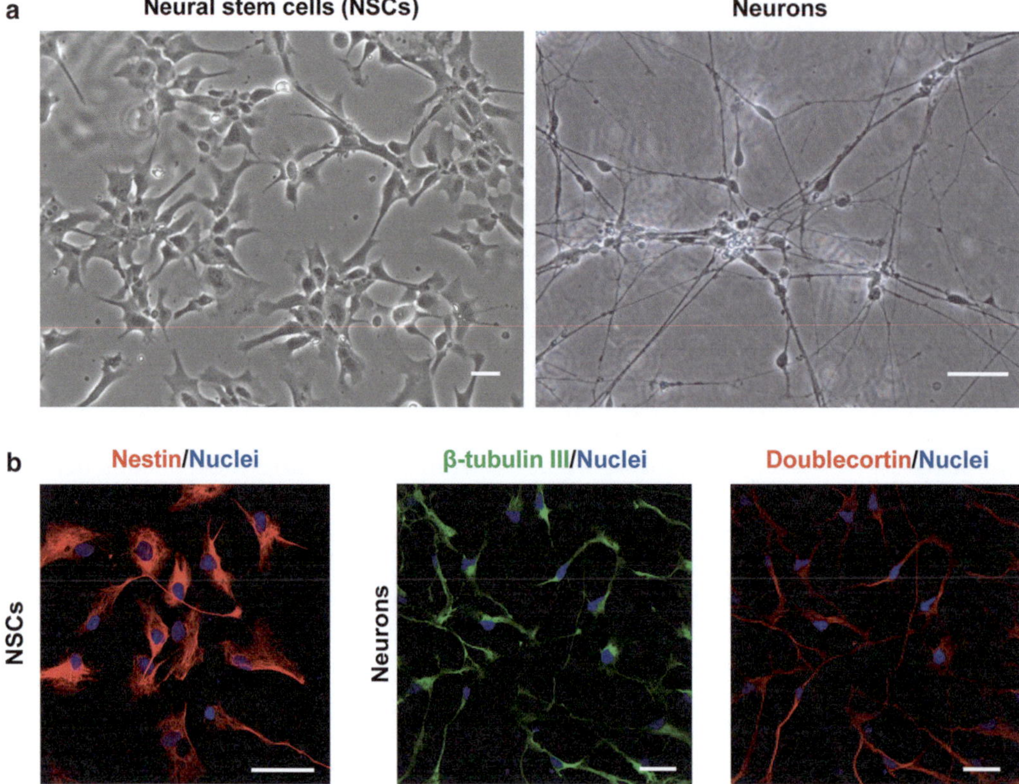

Fig. 1 Characterization of neural stem cells (NSCs) and differentiated neurons. (**a**) Phase-contrast images of cultured NSCs and differentiated neurons. Following the 2-week culture in neuron differentiation medium, NSCs differentiated into neurons with small, round cell bodies and extending long projections. Scale bar = 50 μm. (**b**) Confocal images of NSCs and differentiated neurons upon immunofluorescence staining. NSCs expressed the NSC-specific marker, nestin (*red*). Differentiated neurons stained positive for the neuron-specific marker β-tubulin III (*green*) and immature neuron marker doublecortin (*red*). The *blue dots* are the cell nuclei stained with Hoechst 33342. Scale bar = 20 μm

neurons undergo synaptogenesis and are likely representative of the critical stage of development that is of interest in anesthetic-induced neurotoxicity.

Human neurons are thought to be the most vulnerable to anesthetic administration between the third trimester in utero and the second or third year of life [42]. The exact neuronal markers in humans during this time period are not well understood making it difficult to define this stage of development. Thus, it is difficult to assess the exact maturity level of the NSC-derived neurons. In an attempt to better gauge the maturity level of the NSC-derived neurons, the cells were also immunostained 2 weeks after the initiation of differentiation media for doublecortin, a marker of immature/migrating neurons (Fig. 1b). Based upon the results of this staining, most of the neurons in culture (90–95 %) were positive for this marker of immature neurons. Upon exposure to anesthetics such as propofol and another intravenous anesthetic ketamine, we observed the toxic effect on 2-week-old neurons generated from NSCs [31] as reported in animal studies [43–46], suggesting that this is a valuable model of developing human neurons.

3.2 Analysis of RNA Quality

Determining RNA quality is important prior to performing any downstream analysis such as cDNA conversion and qRT-PCR. An RNA sample of poor quality could compromise the results of the study. The quality of the RNA was assessed using an Epoch nanodrop spectrophotometer using the A260:A280 ratio. An RNA sample is considered "pure" when the A260:A280 is between 1.8 and 2.2. Samples falling out of this range are therefore considered of low quality [47]. The A260:A230 ratio observed in the samples used for our studies was 2.0 ± 0.2, $n = 5$.

3.3 qRT-PCR Analysis of miR-21 Expression

The miR-21 expression was analyzed using qRT-PCR. Specificity of the qRT-PCR primers was confirmed by melting curve analysis. The melting curves of miR-21 and Rnu-6 were sharply defined curves with a single, narrow peak, indicating specificity of the qRT-PCR primers. Figure 2a shows a representative melting curve of miR-21. qRT-PCR analysis demonstrated that propofol downregulated miR-21 after 3 h of exposure and miR-21 was further downregulated after 6 h of propofol exposure (Fig. 2a).

Our recent publication showed that downregulation of miR-21 conferred by propofol exposure played a role in the increased cell death observed in the hESC-derived neurons following propofol administration. We found that overexpression of miR-21 could attenuate the propofol-induced cell death seen in the hESC-derived neurons. Additionally, we found that miR-21 knockdown exacerbated the toxic effects of propofol [31]. These data indicate that miR-21 is playing an important role in the neuronal cell death.

However, the potential role of additional microRNAs in propofol-induced developmental neurotoxicity cannot be excluded. As shown in our previous study [31], with the use of the human miFinder miRNA qRT-PCR arrays, we identified 20 microRNAs that were significantly downregulated following exposure to 6 h of 20 μg/mL propofol when compared to vehicle-treated cells ($P < 0.05$, $n = 4$/group). Of these 20 microRNAs, several were of interest to us based upon their established roles in other diseases or models. For example, the let-7 family has been shown to be highly expressed in the brain and is important in stem cell differentiation and apoptosis [48]. In addition, miRs 9 and 124 have been shown to play a role in neuronal differentiation [49]. The potential roles of these altered microRNAs conferred by propofol are under investigation in our laboratory and much work is needed to fully elucidate the role of microRNAs in anesthetic-induced developmental neurotoxicity.

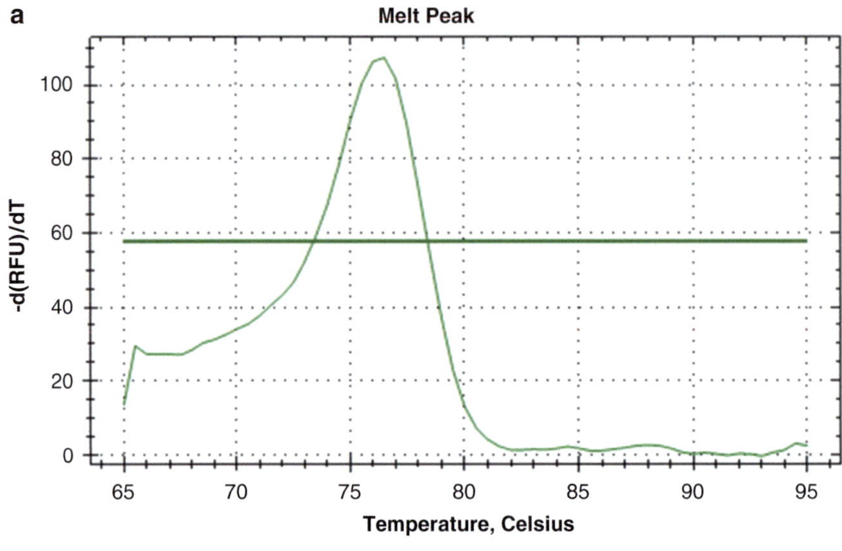

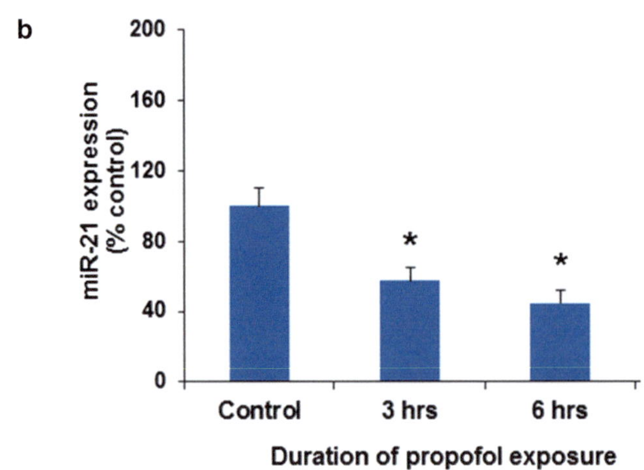

Fig. 2 Time course of miR-21 expression in stem cell-derived human neurons following various lengths of exposure to propofol analyzed using quantitative reverse transcription-PCR (qRT-PCR). (**a**) Representative melting curve of PCR miR-21 product. The melting curve of miR-21 was a single, sharply defined, and narrow peak, indicating the specificity of the qRT-PCR primers. (**b**) miR-21 expression. Two-week-old NSC-derived neurons were exposed to 20 μg/mL propofol for 3 or 6 h. At the end of the exposure, the cells were lysed and miR-21 expression was assessed by qRT-PCR. Propofol significantly decreased miR-21 expression (*$P < 0.05$ vs. control, $n = 6$/group)

3.4 Analysis of Expression of Potential Target Gene of miR-21

miR-21 has many established and predicted targets. Of these targets, programmed cell death protein 4 (PDCD4), Sprouty 1 and 2, and phosphatase and tensin homolog (PTEN) are the most well studied [37]. Protein expression of PDCD4, Sprouty 2, and PTEN was evaluated using Western blot. We found no change in the expression of PTEN in NSC-derived neurons following 6-h-propofol exposure, indicating that PTEN is not involved in this

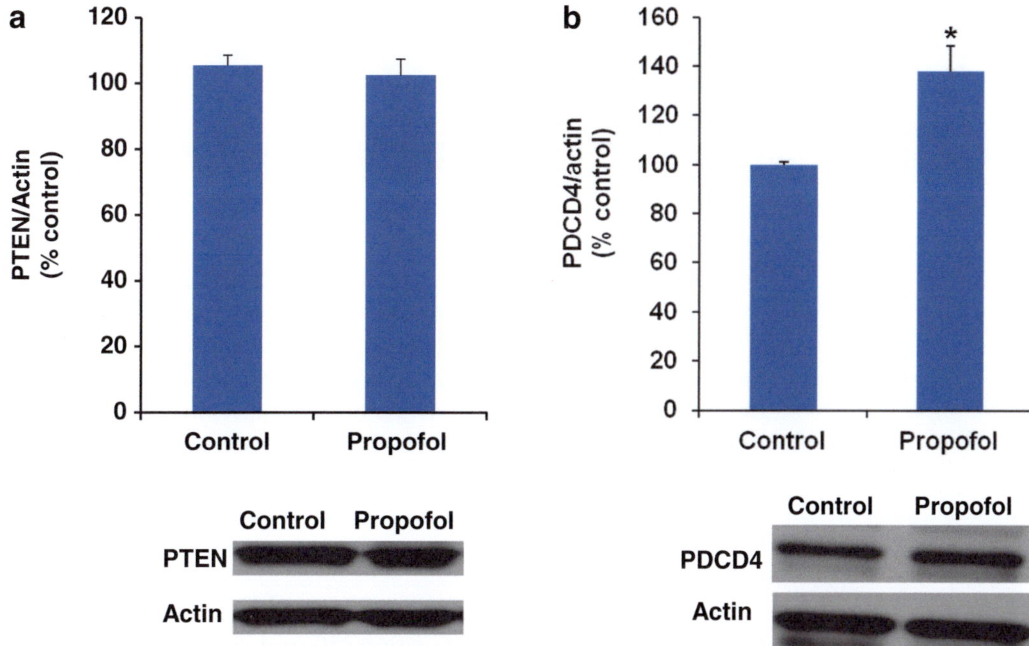

Fig. 3 Propofol exposure increases the expression of PDCD4 but not PTEN in NSC-derived neurons. Following exposure of NSC-derived neurons to 20 μg/mL propofol for 6 h, the expression of PTEN and PDCD4 was assessed by Western blot and normalized to the housekeeping gene, actin. (**a**) PTEN expression. There was no significant change in PTEN expression following propofol exposure. (**b**) PDCD4 expression. PDCD4 expression was significantly elevated in propofol-treated neurons when compared to control-treated cells, suggesting that PDCD4 might be a direct target of miR-21 that is involved in the propofol-induced neurotoxicity in our model (*$P < 0.05$ vs. control or scramble, $n = 3$)

pathway (Fig. 3a). We did find that the expression levels of Sprouty 2 (see publication [24]) and PDCD4 (Fig. 3b) were significantly increased following exposure to propofol, which is consistent with the miR-21 expression data (Fig. 2b) since microRNAs act as negative regulators of their target genes. These data suggest that Sprouty 2 and PDCD4 (but not PTEN) directly targeted by miR-21 might represent possible mechanisms by which propofol induces toxicity in the stem cell-derived human neurons.

Sprouty 2 is a direct target of miR-21 and acts as a negative-feedback regulator of many different receptor tyrosine kinases and inhibits the extracellular-signal-regulated kinase pathway [50]. Overexpression of Sprouty 2 in neurons has been shown to induce cell death and block neurite formation [51]. Although our recent study showed that knockdown of Sprouty 2 partially attenuated the propofol-induced cell death, indicating that Sprouty 2 does play a key role in the propofol-induced neuronal cell death [24], we cannot rule out the role of additional miR-21 targets. Partial attenuation of the propofol-induced cell death by Sprouty 2 knockdown could be due to the incomplete knockdown of Sprouty 2 or could suggest a role for additional miR-21 targets. The expression of PDCD4, another direct target of miR-21, was upregulated following exposure to propofol as assessed by Western blot (Fig. 3b). Future studies may focus on understanding the intricate balance between miR-21 expression and the expression of its many targets and how shifts in that balance might be involved in anesthetic-induced neurotoxicity.

The role of microRNAs in neuronal differentiation, neurological diseases, and neuronal physiology is just beginning to be understood. This work has been done primarily in animals and has been limited by the lack of an appropriate human model. The development of human stem cell-derived neurons allows for new directions of research in these fields using a human model to understand the roles of microRNAs in human brain development and pathology.

Acknowledgement

This work was supported by R01GM112696 from the NIH (to Dr. Xiaowen Bai), by P01GM066730 and R01HL034708 from the NIH, Bethesda, MD, and by FP00003109 from Advancing a Healthier Wisconsin Research and Education Initiative Fund (to Dr. Zeljko J. Bosnjak).

References

1. Zheng H, Dong Y, Xu Z, Crosby G, Culley DJ, Zhang Y et al (2013) Sevoflurane anesthesia in pregnant mice induces neurotoxicity in fetal and offspring mice. Anesthesiology 118 (3):516–526
2. Loepke AW, Soriano SG (2008) An assessment of the effects of general anesthetics on developing brain structure and neurocognitive function. Anesth Analg 106:1681–1707
3. Jevtovic-Todorovic V (2011) Pediatric anesthesia neurotoxicity: an overview of the 2011 SmartTots panel. Anesth Analg 113:965–968
4. Liu F, Paule MG, Ali S, Wang C (2011) Ketamine-induced neurotoxicity and changes in gene expression in the developing rat brain. Curr Neuropharmacol 9:256–261
5. Lemkuil BP, Head BP, Pearn ML, Patel HH, Drummond JC, Patel PM (2011) Isoflurane neurotoxicity is mediated by p75NTR-RhoA activation and actin depolymerization. Anesthesiology 114:49–57
6. Sun L (2010) Early childhood general anaesthesia exposure and neurocognitive development. Br J Anaesth 105(Suppl 1):i61–i68
7. Stratmann G, Sall JW, May LD, Bell JS, Magnusson KR, Rau V et al (2009) Isoflurane differentially affects neurogenesis and long-term neurocognitive function in 60-day-old and 7-day-old rats. Anesthesiology 110:834–848
8. Samuelsen GB, Larsen KB, Bogdanovic N, Laursen H, Graem N, Larsen JF et al (2003) The changing number of cells in the human fetal forebrain and its subdivisions: a stereological analysis. Cereb Cortex 13:115–122
9. Dekaban AS (1978) Changes in brain weights during the span of human life: relation of brain weights to body heights and body weights. Ann Neurol 4:345–356
10. Davidson AJ, McCann ME, Morton NS, Myles PS (2008) Anesthesia and outcome after neonatal surgery: the role for randomized trials. Anesthesiology 109:941–944
11. Hansen TG, Flick R (2009) Anesthetic effects on the developing brain: insights from epidemiology. Anesthesiology 110:1–3
12. Istaphanous GK, Howard J, Nan X, Hughes EA, McCann JC, McAuliffe JJ et al (2011) Comparison of the neuroapoptotic properties of equipotent anesthetic concentrations of

desflurane, isoflurane, or sevoflurane in neonatal mice. Anesthesiology 114:578–587

13. Zhang Y, Dong Y, Wu X, Lu Y, Xu Z, Knapp A et al (2010) The mitochondrial pathway of anesthetic isoflurane-induced apoptosis. J Biol Chem 285:4025–4037
14. Sun LS, Li G, Dimaggio C, Byrne M, Rauh V, Brooks-Gunn J et al (2008) Anesthesia and neurodevelopment in children: time for an answer? Anesthesiology 109:757–761
15. Thomson JA, Itskovitz-Eldor J, Shapiro SS, Waknitz MA, Swiergiel JJ, Marshall VS et al (1998) Embryonic stem cell lines derived from human blastocysts. Science 282:1145–1147
16. Bartel DP (2009) MicroRNAs: target recognition and regulatory functions. Cell 136:215–233
17. Bartel DP (2004) MicroRNAs: genomics, biogenesis, mechanism, and function. Cell 116:281–297
18. Ivanovska I, Ball AS, Diaz RL, Magnus JF, Kibukawa M, Schelter JM et al (2008) MicroRNAs in the miR-106b family regulate p21/CDKN1A and promote cell cycle progression. Mol Cell Biol 28:2167–2174
19. Lee RC, Feinbaum RL, Ambros V (1993) The C. elegans heterochronic gene lin-4 encodes small RNAs with antisense complementarity to lin-14. Cell 75:843–854
20. Lee CT, Risom T, Strauss WM (2007) Evolutionary conservation of microRNA regulatory circuits: an examination of microRNA gene complexity and conserved microRNA-target interactions through metazoan phylogeny. DNA Cell Biol 26:209–218
21. Shukla GC, Singh J, Barik S (2011) MicroRNAs: processing, maturation, target recognition and regulatory functions. Mol Cell Pharmacol 3:83–92
22. Sathyan P, Golden HB, Miranda RC (2007) Competing interactions between micro-RNAs determine neural progenitor survival and proliferation after ethanol exposure: evidence from an ex vivo model of the fetal cerebral cortical neuroepithelium. J Neurosci 27:8546–8557
23. Buller B, Liu X, Wang X, Zhang RL, Zhang L, Hozeska-Solgot A et al (2010) MicroRNA-21 protects neurons from ischemic death. FEBS J 277:4299–4307
24. Twaroski DM, Yan Y, Olson JM, Bosnjak ZJ, Bai X (2014) Down-regulation of MicroRNA-21 is involved in the propofol-induced neurotoxicity observed in human stem cell-derived neurons. Anesthesiology 121(4):786–800
25. Lagos-Quintana M, Rauhut R, Lendeckel W, Tuschl T (2001) Identification of novel genes coding for small expressed RNAs. Science 294:853–858
26. Ge XT, Lei P, Wang HC, Zhang AL, Han ZL, Chen X et al (2014) miR-21 improves the neurological outcome after traumatic brain injury in rats. Sci Rep 4:6718
27. Zhang Y, Tian J, Chen S, Zhang X, Cao SE (2014) Role of miR-34c in ketamine-induced neurotoxicity in neonatal mice hippocampus. Cell Biol Int. doi:10.1002/cbin.10349
28. Xu H, Zhang J, Zhou W, Feng Y, Teng S, Song X (2015) The role of miR-124 in modulating hippocampal neurotoxicity induced by ketamine anesthesia. Int J Neurosci 125:213–220
29. Bosnjak ZJ, Yan Y, Canfield S, Muravyeva MY, Kikuchi C, Wells CW et al (2012) Ketamine induces toxicity in human neurons differentiated from embryonic stem cells via mitochondrial apoptosis pathway. Curr Drug Saf 7:106–119
30. Liao H, Huang W, Schachner M, Guan Y, Guo J, Yan J et al (2008) Beta 1 integrin-mediated effects of tenascin-R domains EGFL and FN6-8 on neural stem/progenitor cell proliferation and differentiation in vitro. J Biol Chem 283:27927–27936
31. Twaroski DM, Yan Y, Olson JM, Bosnjak ZJ, Bai X (2014) Down-regulation of microRNA-21 is involved in the propofol-induced neurotoxicity observed in human stem cell-derived neurons. Anesthesiology 121:786–800
32. Vutskits L, Gascon E, Tassonyi E, Kiss JZ (2005) Clinically relevant concentrations of propofol but not midazolam alter in vitro dendritic development of isolated gamma-aminobutyric acid-positive interneurons. Anesthesiology 102:970–976
33. Chung HG, Myung SA, Son HS, Kim YH, Namgung J, Cho ML et al (2013) In vitro effect of clinical propofol concentrations on platelet aggregation. Artif Organs 37:E51–E55
34. Ludbrook GL, Visco E, Lam AM (2002) Propofol: relation between brain concentrations, electroencephalogram, middle cerebral artery blood flow velocity, and cerebral oxygen extraction during induction of anesthesia. Anesthesiology 97:1363–1370
35. Costela JL, Jimenez R, Calvo R, Suarez E, Carlos R (1996) Serum protein binding of propofol in patients with renal failure or hepatic cirrhosis. Acta Anaesthesiol Scand 40:741–745
36. Cheng Y, Liu X, Zhang S, Lin Y, Yang J, Zhang C (2009) MicroRNA-21 protects against the H(2)O(2)-induced injury on cardiac myocytes via its target gene PDCD4. J Mol Cell Cardiol 47:5–14
37. Buscaglia LE, Li Y (2011) Apoptosis and the target genes of microRNA-21. Chin J Cancer 30:371–380

38. Meng F, Henson R, Wehbe-Janek H, Ghoshal K, Jacob ST, Patel T (2007) MicroRNA-21 regulates expression of the PTEN tumor suppressor gene in human hepatocellular cancer. Gastroenterology 133:647–658
39. Jevtovic-Todorovic V (2012) Developmental synaptogenesis and general anesthesia: a kiss of death? Curr Pharm Des 18:6225–6231
40. Lunardi N, Ori C, Erisir A, Jevtovic-Todorovic V (2010) General anesthesia causes long-lasting disturbances in the ultrastructural properties of developing synapses in young rats. Neurotox Res 17:179–188
41. Bai X, Yan Y, Canfield S, Muravyeva MY, Kikuchi C, Zaja I et al (2013) Ketamine enhances human neural stem cell proliferation and induces neuronal apoptosis via reactive oxygen species-mediated mitochondrial pathway. Anesth Analg 116:869–880
42. Dobbing J, Sands J (1979) Comparative aspects of the brain growth spurt. Early Hum Dev 3:79–83
43. Creeley C, Dikranian K, Dissen G, Martin L, Olney J, Brambrink A (2013) Propofol-induced apoptosis of neurones and oligodendrocytes in fetal and neonatal rhesus macaque brain. Br J Anaesth 110(Suppl 1):i29–i38
44. Milanovic D, Popic J, Pesic V, Loncarevic-Vasiljkovic N, Kanazir S, Jevtovic-Todorovic V et al (2010) Regional and temporal profiles of calpain and caspase-3 activities in postnatal rat brain following repeated propofol administration. Dev Neurosci 32:288–301
45. Dong C, Anand KJ (2013) Developmental neurotoxicity of ketamine in pediatric clinical use. Toxicol Lett 220:53–60
46. Scallet AC, Schmued LC, Slikker W Jr, Grunberg N, Faustino PJ, Davis H et al (2004) Developmental neurotoxicity of ketamine: morphometric confirmation, exposure parameters, and multiple fluorescent labeling of apoptotic neurons. Toxicol Sci 81:364–370
47. Lam B, Simkin M, Rghei N, Haj-Ahmad Y (2012) Revised guidelines for rna quality assessment for diverse biological sample input. https://norgenbiotek.com/sites/default/files/resources/ffpe_rna_purification_kit_revised_guidelines_for_rna_quality_assessment_for_diverse_biological_sample_input_application_notes_63.pdf
48. Roush S, Slack FJ (2008) The let-7 family of microRNAs. Trends Cell Biol 18:505–516
49. Roese-Koerner B, Stappert L, Koch P, Brustle O, Borghese L (2013) Pluripotent stem cell-derived somatic stem cells as tool to study the role of microRNAs in early human neural development. Curr Mol Med 13:707–722
50. Sayed D, Rane S, Lypowy J, He M, Chen IY, Vashistha H et al (2008) MicroRNA-21 targets Sprouty2 and promotes cellular outgrowths. Mol Biol Cell 19:3272–3282
51. Gross I, Armant O, Benosman S, de Aguilar JL, Freund JN, Kedinger M et al (2007) Sprouty2 inhibits BDNF-induced signaling and modulates neuronal differentiation and survival. Cell Death Differ 14:1802–1812

Neuromethods (2017) 128: 43–57
DOI 10.1007/7657_2017_1

Published online: 21 April 2017

MicroRNA Expression Profiling by PCR Array in 2D and 3D Differentiated Neural Culture Systems and Target Validation

Lara Stevanato, Caroline Hicks, Lavaniya Thanabalasundaram, and John D. Sinden

Abstract

MicroRNAs (miRNAs) have been proven to regulate gene expression at post-transcriptional level and are emerging as strong mediators in neural fate determination (Ambros, Nature 431(7006):350–355, 2004). Here, we evaluated appropriate 3 three dimensional (3D) substrates to differentiate human neural stem cells (hNSCs). We identified and quantified hNSC miRNA contents by PCR array. By using computational algorithms we predicted miRNA target mRNA which correlates with hNSC differentiation and performed target validation by transfection of 3 prime untranslated regions (3′UTR) dual reporter plasmids and dual luciferase assay. Despite the inherent differences between cultures, we were able to consistently show that 3D topography promotes differentiation of hNSCs through modulation of miRNAs associated with cell proliferation and maintenance of stemness.

Keywords: Clinical grade neural stem cells, In vitro differentiation, miRNA profiling and effects, miRNA target validation, Three dimensional culture

Abbreviations

1 W	1 Week
2D	Two dimensional
3 W	3 Weeks
3′UTR	3 Prime untranslated region
3D	Three dimensional
4-OHT	4-Hydroxytamoxifen
bFGF	Basic fibroblast growth factor
DCX	Doublecortin
EGF	Epidermal growth factor receptor
GALC	Galactosylceramidase
GFAP	Glial fibrillary acidic protein
HNSC	Human neural stem cell
Hsa-miR	Human miRNA
ICC	Immunocytochemistry
MAP2	Microtubule-associated protein 2
miRNA	MicroRNA
mRNA	Messenger RNA

QRT-PCR	Real-time reverse transcription PCR
S100B	S100 calcium binding protein B
TUBB3	Tubulin, beta 3 class III

1 Introduction

MiRNAs are short 20–22 nucleotide RNA molecules, they are tissue-specifically expressed and developmentally regulated [1]. They function as negative regulators of gene expression in a variety of eukaryotic organisms [2] and are key post-transcriptional players in stem cell self-renewal and fate determination [3]. They do not encode proteins, but bind the 3 prime untranslated region (3′UTR) of mRNAs and regulate their stability and translation into proteins. Each miRNA can regulate one or more mRNA transcripts, and conversely a given mRNA can be regulated by one or more miRNA. Therefore, although they are well-characterized, the complex part played by each miRNA has yet to be completely defined.

The complexity of the in vivo environment is not mimicked in standard 2D in vitro differentiation protocols and consequently the induction and regulation of hNSC differentiation may not be optimal in these systems. 3D microenvironments, tissue-specific architecture, mechanical and biochemical cues, extracellular ligands, including many types of collagens, laminin, and other matrix proteins (extracellular matrix, ECM) are pivotal for in vivo cell differentiation. Previous studies have shown that materials mimicking ECMs and their geometry influence cell phenotype and fate [4–11]. Environmental cues and in vivo 3D environments can be critical for cellular simulation of normal cell morphology, proliferation, differentiation, and migration, and can be mimicked using synthetically engineered 3D culture systems.

1.1 Methods to Study Cell Differentiation on 3D Cultures

Scaffolds can be synthesized to support 3D cell growth and have direct applications in tissue engineering and regenerative medicine. Many different types of scaffolds have been developed as collagen-based hydrogels [12], poly(L-lactide) (PLLA) [13], and polystyrene [14] scaffolds; poly(dl-lactic acid-co-glycolic acid) (PLGA) [15] microspheres, and native tissue scaffolds [16]. However, there is no universal scaffold. The type of scaffold to be used depends on its proposed function and desired characteristics. A wide range of characteristics determining design criteria for 3D scaffolds include the following: biomaterial; biocompatibility; biodegradability; porosity, pore size; geometry; co-culture of cells; shape and size; inter-connectivity; orientation; mechanical properties. In here we report the investigation of hNSC neural and glial differentiation on commercially available 3D scaffolds. Among those tested, Alvetex® (Amsbio, UK) was selected to further perform miRNA profiling characterization and target validation. Alvetex® is a highly porous polystyrene scaffold engineered with a well-defined and uniform architecture into a 200 μm thick membrane which provides a 3D space into which cells can occupy and differentiate [14]. The scaffold is formed by polymerization in a biphasic emulsion, consisting of an aqueous and a non-aqueous monomer/surfactant phase, termed a high internal phase emulsion (HIPE) [17, 18]. The resulting polymer (poly-HIPE) consists of a relatively homogeneous porous network of voids, linked by inter-connecting pores. Moreover, such scaffolds have been proven amenable for 3D in vitro cell cultures [14].

1.2 Methods to Study miRNA Profiling and Target Validation

There are several methods to profile and to quantify miRNA expression in cells and tissues. They range from measuring miRNAs by conventional microarrays, quantitative PCR, and next generation sequencing. Profiling of specific under or over-expressed miRNAs can be used to identify key miRNAs involved in cell biological processes such as differentiation. Identification of miRNA/mRNA putative target is fundamental to unravel miRNA functionality. Several algorithms have been developed to help predict miRNA target mRNA [19, 20]. Experimental procedures should be implemented to demonstrate that a given mRNA is a target of a specific miRNA. Direct miRNA/mRNA targets can be validated using 3′UTR luciferase reporter assays [21]. We utilized quantitative PCR arrays for the profiling of miRNAs involved in stem cells and development pathways and validated miRNA/mRNA targets.

In here, as exemplification of the proposed methods, we report the results of our recent study [21].

2 Materials

2.1 HNSC Derivation, Culture, Differentiation, and Immunocytochemistry

We used cultures of CTX0E03, a fully manufactured conditionally immortalized hNSC line, originally derived from ethically sourced human fetal brain cortical tissue of 12 weeks gestation as described in [22] to quantify and image hNSC differentiation. HNSCs were cultured in a defined medium (RMM) supplemented with epidermal growth factor (EGF, 20 ng/ml, Peprotech), basic fibroblast growth factor (bFGF, 10 ng/ml, Peprotech), and 4-hydroxytamoxifen (4-OHT, 10 mM, Sigma) on laminin (20 μg/ml, AMS Biotech) coated vessels and incubated in a humidified atmosphere containing 5% CO_2. The RMM is prepared by combining DMEM: F12 (Invitrogen) with the following reagents: 0.03% human albumin solution (Baxter health care limited), 5 μg/ml human recombinant transferrin (Sigma), 16.2 μg/ml Putrescine DiHCl (Sigma), 5 μg/ml human recombinant Insulin (Sigma), 60 ng/ml progesterone (Sigma), 2 mM L-glutamine (Invitrogen), and 40 ng/ml sodium selenite (Sigma). All medium components are prepared as individual aliquots with a shelf life up to 3 months when stored under appropriate conditions. Once the medium is formulated it is given a 2 week expiry date.

For differentiation assays a single cell suspension of hNSCs, achieved by trypsinization (TrypZean/EDTA, Lonza), followed by trypsin inhibition using defined trypsin inhibitor (Invitrogen) and centrifugation, was seeded either on laminin coated standard cell-treated plastic vessels (BD Biosciences) or on 3D scaffolds (Alvetex®, Amsbio, and Electrospinning) and grown for 1 and 3 weeks in RMM without growth factors (GFs) and 4-OHT. HNSC differentiation on different scaffolds was evaluated by immunocytochemistry (ICC). After culture, HNSCs were fixed with 4% PFA (Pioneer Research) and incubated with a cocktail of anti-TUBB3 (1:1,000, Sigma) and anti-GFAP (1:2,500, Dako) antibodies diluted in 1× PBS containing 1% normal goat serum (Vector laboratories) and detected with anti-mouse conjugated Alexa Fluor 488 (Invitrogen, 1:2,000) and anti-rabbit conjugated Alexa Fluor 568 (Invitrogen, 1:2,500) secondary antibodies for TUBB3 and GFAP, respectively. HNSCs nuclei were counterstained with Hoechst 33342 (Sigma, 1 μM). Images were captured using an inverted microscope (Olympus IX70 fluorescent microscope) and Image-Pro Plus 7 (Media Cybernetics) imaging software.

2.2 Real-Time RT-PCR (qRT-PCR) Analysis of mRNA, Total RNA Extraction, and cDNA Preparation

To investigate hNSC differentiation by qRT-PCR, total RNA was isolated using miRNeasy (Qiagen). Proteins and cell debris were removed by the addition of chloroform (200 μl, Sigma). Upper aqueous phases (500 μl) were combined with 100% ethanol (750 μl, Merck Millipore) and uploaded on RNeasy® Mini columns (Qiagen). Total RNA concentration and 260/280 ratio were measured using a NanoDrop Lite spectrophotometer (ThermoScientific).

A minimum of 250 ng of total RNA was reverse-transcribed into first-strand cDNA using a mix of random primers (100 ng/μl, Invitrogen), poly-dT (50 ng/μl, Invitrogen), 1 μl of dNTP mix (10 mM each, Invitrogen), and superscript II reverse transcriptase mix (4 μl 5× first-strand buffer, 2 μl 0.1 M DTT, 1 μl RNasin, Promega, 40 units/μl, and 1 μl superscript II RT, Invitrogen).

Two μl of cDNA, 5 μl lightcycler 480 probe master (Roche), 0.2 μl of human universal probe library (UPL, Roche), 1 μl of each 0.4 μM primer set (Sigma), and 1.8 μl RNase-free water were combined to give a final volume of 10 μl for each qRT-PCR reaction. The primer sequences and UPL used for each gene are listed in Table 1. Roche LC480 real-time cycler was programmed accordingly: 1 cycle at 95 °C for 10 min, and 35 cycles for 10 s at 95 °C, and 30 s at 55 °C, and 72 °C set as single (fluorescence data collection). Two housekeeping genes, ATP5B and YWHAZ (PrimerDesign) were used for the gene normalization.

Table 1
Selected neural and glial markers

Gene	Access N	Forward primer	Reverse primer	UPL
TUBB3	NM_006086	gcaactacgtgggcgact	cgaggcacgtacttgtgaga	78
DCX	NM_000555.2	gtggaggctggtaaagagca	aggcccaagcataaggaaat	6
MAP2	NM_031845.2	cgaactttatattttaccacttccttg	ccgttcatctgccattcttc	2
GFAP	NM_002055.3	ccagttgcagtccttgacct	tctccagggactcgttcgt	88
S100B	NM_006272.2	caggatccttgcctccaac	ctcagagcccccggtagt	67
GALC	NM_001037525.1	tggtgcctctttgcatatttta	atgtgggagggctcagtg	9
SOX5	NM_006940.4	tttacctcaggagtttgaaagga	gcttgtcaccatggctacct	38
FOXN3	NM_005197.3	cattaagaggtgtggcgttttt	gacacatgaaccgccactt	3
NR4A3	NM_173199	tctcagtgttggaatggtaaaaga	ggtttggaaggcagacgac	52
DUSP10	NM_007207.4	tgaatgtgcgagtccatagc	tggcaattcaagaagaactcaa	22
EIF4G3	NM_003760.4	attctcaaaacttaaattcaagaagga	tttcttccatgtctttggtacagt	33

Gene, access number (N), F & R primers and UPL

2.3 QRT-PCR Analysis of miRNA Using Stem Cells and Developmental Pathways Focused miRNA PCR Array

To investigate differentiated hNSC miRNA profiling, we used stem cells and developmental pathways focused miRNA PCR arrays. For each array a minimum of 250 ng total RNA was retro-transcribed using miScript II RT Kit (Qiagen) and combined with 2 μl of 10× miScript nucleics mix, 4 μl of 5× miScript HiSpec buffer, 2 μl of superscript II reverse transcriptase, and a variable amount of RNase-free water to give a final volume of 20 μl. The stem cells and developmental pathways focused miRNA PCR array was (Qiagen, SABiosciences) carried out by preparing a PCR component mix in a 15 ml tube by adding 1,375 μl of 2× RT^2 SYBR green master mix (Qiagen), 100 μl miRNA cDNA preparation, 275 μl of 10× miScript Universal Primer, and 1,000 μl of RNase-free water (total volume 2,750 μl). 25 μl of PCR component mix was added to each well of the PCR array using a multi-channel pipettor. A Roche LC480 real-time cycler was programmed accordingly: 1 cycle at 95 °C for 10 min, and 45 cycles for 15 s at 95 °C, and 1 min at 60 °C set as single (fluorescence data collection).

2.4 Computational Target Gene Predictions, Validation by qRT-PCR, and Pathway Analysis

DIANA-microT 3.0 [23, 24], PicTar [25, 26], and TargetScan [20, 27] algorithms were used to identify miRNA target prediction mRNA.

2.5 Reporter Plasmid Transfection and Dual Luciferase Assay

Transient transfections of HeLa cells were performed using Lipofectamine 2000 (Invitrogen). DNA-miRNA-Lipofectamine® 2000 complexes must be made in serum-free medium such as Opti-MEM® (Invitrogen). For the luciferase assay, HeLa were seeded at a density of 10^5 cells/well in 24-well plates. For each well plasmid DNA-lipid complexes were prepared by combining 1.5 μl of Lipofectamine 2000 with 100 ng of MiTarget™ MicroRNA 3′UTR Target Clone HmiT019538-MT01 (GeneCopoeia, NR4A3 3′ UTR) or HmiT017632-MT01 (GeneCopoeia, SOX5 3′ UTR), and 20 nM of miRNA mimics (hsa-miR-96-5p for SOX5 3′UTR, and hsa-miR-7-5p, and hsa-miR-17-5p for NR4A3 3′UTR, respectively). Control wells were transfected with either reporter plasmids or allstars negative control siRNA AF 488 (Qiagen). HmiT019538-MT0 and HmiT017632-MT01 plasmids expressed both firefly and renilla luciferase. Firefly and renilla luciferase activities were using the Luc-Pair miR luciferase assay (GeneCopoeia) and a GloMax™ 96 Microplate Luminometer (Promega).

3 Methods

3.1 Identification of a Suitable 3D Scaffolds and Quantification of Differentiation by qRT-PCR Molecular Analysis

In vitro hNSC differentiation is routinely conducted on two dimensional (2D) substrates, which does not accurately mimic the 3D environment of living tissues. In vivo, the 3D structure provides optimal conditions for cell growth and function. We have characterized the experimental conditions required to allow hNSCs attachment, and differentiation on 3D cultures. We conducted our evaluation on commercially available scaffolds as presented in Fig. 1 and we selected to use polystyrene scaffolds (Alvetex®, Amsbio) for further investigation. Evaluation of optimal cell seeding and requirement of a substrate coating, in our case laminin, is a fundamental feature for the development of a robust in vitro differentiation assay. The optimal hNSC seeding concentration was found to be 2.5×10^5 cells/cm^2.

Immunocytochemistry (ICC) is a conventionally established technique used for the investigation and quantification of cell differentiation [22]. The main disadvantages of this method are difficulties in quantification and sampling error, only limited representative image fields can be captured, furthermore standardization and interpretation of results can be operator biased. We propose the use of qRT-PCR as an alternative analytical method to investigate hNSC neuronal and glial differentiation. Measuring transcript abundance by qRT-PCR has become the method of choice due to its high sensitivity, specificity, and broad quantification range for high-throughput and accurate expression profiling of selected genes [28]. QRT-PCR outputs are referred as crossing point (cp) or cycling thresholds (ct) values, depending on the platform use and on the analysis software, in our case Roche LC480.

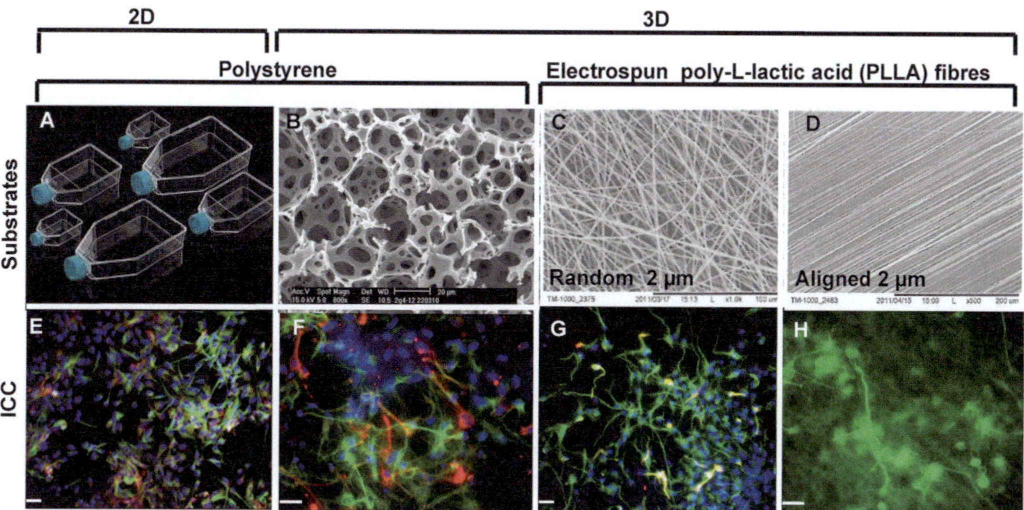

Fig. 1 Evaluation of hNSC differentiation on 3D scaffolds. Representative images of substrate types (polystyrene and PLLA) and hNSCs grown on 2D (**a**, **e**), or on 3D cultures: polystyrene scaffold (**b**, **f**) or PLLA (random, **c**, **g**, or aligned **d**, **h**) revealing the differences between 2D and 3D differentiation, especially of axonal outgrowth. (**e**–**h**) Representative images showing TUBB3 (*green*), GFAP (*red*), and nuclei (*blue*) staining, bar 50 μm

These values can be exported to perform relative quantification, based on a comparative ct method. Changes in gene of interest (Table 1, Fig. 2) expression in a given sample (e.g., differentiated) can be compared to another reference sample (such as an undifferentiated control sample), calculated, and indicated as Δct of gene of interest. Normalization, required to take in account difference in initial sample concentration input in the investigation, is indicated as Δct of reference gene. Relative quantification is calculated using the formula: $2^{-\Delta\Delta ct}$ [29] and takes in account both Δct of gene of interest and reference gene (ΔΔct). Our quantification of neuronal and glial 3D hNSC differentiation is presented in Fig. 2.

3.2 Human Cell Differentiation and Development miScript miRNA PCR Array Profiles

Highly coordinated programmes of gene expression are involved in the development of the nervous system. Lately, it has become clear that gene expression can also be modulated by several classes of small RNAs. To investigate the effect of 3D induced differentiation on hNSC miRNA expression we used a PCR array. Compared with standard chip microarrays PCR arrays offer the advantages of a direct quantification of miRNA of interest. Although the PCR array is limited in terms of total number of miRNA per array it can be tailored to specifically include miRNAs of interest and maximizes the likelihood of discovering miRNAs whose expression patterns correlate with the biological phenotypes under study, in our case in stem cell development [30]. The array consists of 84 stem cell focused miRNAs plus a set of controls that enables relative quantification and assessment of reverse transcription and PCR performance. The geo-mean of controls was used for data analysis

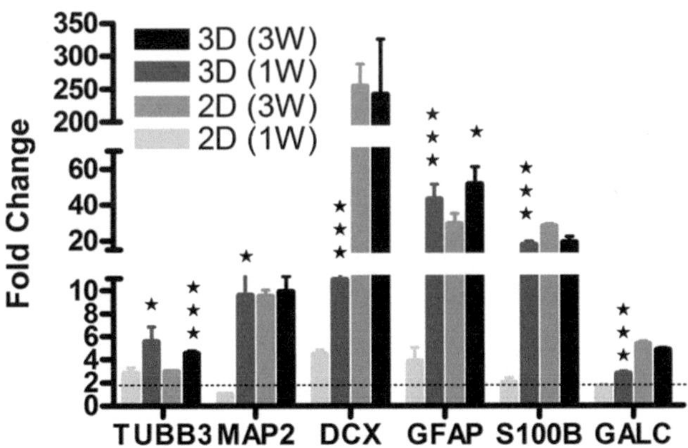

Fig. 2 Quantification of hNSC differentiation. QRT-PCR molecular analysis for β3-tubulin [TUBB3], a well-established neuron-specific marker expressed by neuronal precursors [31]; doublecortin [DCX], expressed in developing neurons, and increasingly used as a marker for neurogenesis [32], microtubule-associated protein 2 [MAP2], involved in microtubule assembly, which is an essential step in neuritogenesis and glial markers: glial fibrillary acidic protein [GFAP], a well-recognized astrocyte marker; S100β, a glial-specific marker and expressed primarily by astrocytes; galactocerebroside [GALC], expressed in differentiating oligodendrocyte precursor cells [33]. Data from 2D to 3D cultures at 1 (1W) and 3 week (3W) post-seeding were compared. GALC, GFAP, TUBB3, S100B, DCX, and MAP2 mRNA expression level analysis showed a significant increase in all markers in 3D hNSCs cultures grown for 1W compared with 2D differentiated hNSCs at the same time point. Furthermore, TUBB3 and GFAP were significantly enhanced in 3D cultures compared with those differentiated on the 2D substrate at the 3W time point. Statistical analysis showed significant difference between 2D and 3D samples at the same time point; ± SDMs, * $p < 0.05$, ** $p < 0.01$, *** $p < 0.005$, Student's *t*-test [21]

based on the $2^{-\Delta\Delta ct}$ method. Three biological replicates were assessed for each condition. Ct values for all wells were exported to a blank Excel spreadsheet and analysed using the SABiosciences PCR Array Data Analysis web-based software. Software is available at www.SABiosciences.com/pcrarraydataanalysis. The miRNA profiling identified several significantly down-regulated miRNAs (Fig. 3); to facilitate data interpretation they were grouped by functionality (Table 3), time, and type of culture conditions.

3.3 MiRNA Target Prediction Analysis, Validated by qRT-PCR, Dual Luciferase

The most significantly down-regulated miRNAs (Table 2) were selected and analysed to assess their putative target mRNAs with the intent of determining their functionality and biological interpretation using available online algorithms. Computational algorithms are commonly the major tool in predicting miRNA targets [31]. These approaches are mainly based on the identification of complementary elements in the 3′UTR with the seed sequence of

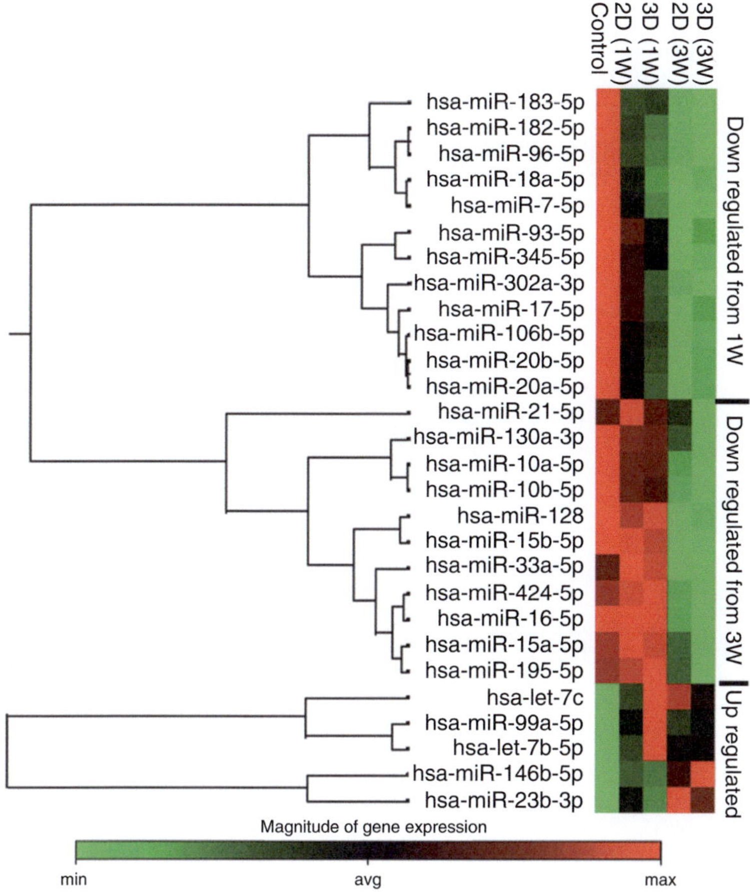

Fig. 3 Human cell differentiation and development of miScript miRNA PCR array profiles. Group clustergram analysis of miRNA differentially regulated in control (undifferentiated hNSCs) and hNSCs seeded on 2D and 3D substrates and differentiated for 1W and 3W [21]. Time dependent down-regulation was observed. At both time points a significant difference was present in 2D compared with 3D substrate types regarding miRNAs belonging to cluster miR-17-92 [39], miR-96-182 [40] (regulators of cell proliferation), and hsa-miR-302 (maintenance of stemness, Table 2). Overall the 2D and 3D miRNA profiles were very similar in terms of miRNA types, however, the degree and timing of miRNA differential regulation was significantly different for miRNAs involved in both maintenance of stemness

the miRNA and its phylogenetic conservation of the complementary sequences of orthologous genes. However, evidence suggests that perfect seed pairing may not necessarily be a reliable prediction for miRNA interactions [32], which may explain why some predicted target sites are incorrect (Fig. 4a). Hence, targets for miRNAs remain to be identified or verified experimentally. Although this analysis is constrained in terms of predictive ability, it is the only tool available for putative miRNA targets, and the output is valuable for proposing hypothetical associations between miRNAs, targeted

Table 2
Statistical analysis performed on miRNA PCR arrays

Mature ID	*P*-values (comparing to control group)				*P*-values (comparing to 2D versus 3D)		Regulation
	2D (1W)	3D (1W)	2D (3W)	3D (3W)	(1W)	(3W)	
hsa-miR-302a-3p	*	***	***	***	***	**	Down
hsa-miR-10a-5p	NS	NS	***	***	***	*	
hsa-miR-10b-5p	NS	*	***	***	NS	NS	
hsa-miR-96-5p	***	***	***	***	***	**	
hsa-miR-183-5p	***	***	***	***	***	***	
hsa-miR-182-5p	***	***	***	***	*	NS	
hsa-miR-7-5p	***	***	***	***	**	NS	
hsa-miR-17-5p	***	***	***	***	***	***	
hsa-miR-18a-5p	***	***	***	***	***	***	
hsa-miR-20a-5p	**	***	***	***	***	***	
hsa-miR-20b-5p	***	***	***	***	*	***	
hsa-miR-93-5p	***	***	***	***	***	**	
hsa-miR-106b-5p	***	***	***	***	***	**	
hsa-miR-15a-5p	NS	NS	***	*	NS	NS	
hsa-miR-15b-5p	NS	NS	***	***	NS	NS	
hsa-miR-16-5p	NS	NS	***	**	NS	NS	
hsa-miR-195-5p	NS	*	***	**	NS	NS	
hsa-miR-21-5p	*	NS	NS	**	***	**	
hsa-miR-33a-5p	NS	NS	***	**	NS	NS	
hsa-miR-128	NS	NS	*	*	NS	NS	
hsa-miR-424-5p	NS	NS	**	***	***	NS	
hsa-miR-130a-3p	NS	*	***	*	NS	NS	
hsa-miR-345	*	***	***	***	NS	NS	
hsa-let-7b	***	***	***	***	***	NS	Up
hsa-let-7c	***	***	***	***	***	***	
hsa-miR-146b-5p	***	*	***	***	***	NS	
hsa-miR-23b	***	***	***	***	***	NS	
hsa-miR-99a	***	***	***	*	***	NS	

Data were analysed by Student's *t*-test, * $p < 0.05$, ** $p < 0.001$, *** $p < 0.005$, and non-significant (NS)

pathways, and biological functions. Since different algorithms often yield different outputs for putative target mRNAs we utilized three online available algorithms: Diana Lab, PicTar, and TargetScan. Overall, the number of resulting target sites was considered high (Table 4). To overcome this issue we selected only putative miRNA target genes that were previously referenced to brain development and function or target sites of multiple miRNAs. The validation of predicted target mRNAs was performed by qRT-PCR (Fig. 4a). We selected two mRNAs for further functional validation of miRNA/mRNA interaction (Fig. 4b). The rationale for using this assay is that the binding of a given miRNA to its specific 3′UTR mRNA target site will repress reporter protein production thereby reducing activity/expression that can be measured and compared

Table 3
List of function of down-regulated miRNAs

Mature ID	Family/cluster	Function	References
hsa-miR-302a-3p hsa-miR-10a-5p hsa-miR-10b-5p hsa-miR-424-5p hsa-miR-130a-3p hsa-miR-345		**Maintenance and regulation of pluripotency**	[33, 34]
hsa-miR-96-5p hsa-miR-183-5p hsa-miR-182-5p hsa-miR-7-5p hsa-miR-128	**miR-96-182**	**Neuronal lineage-specification and differentiation**	[35, 36]
hsa-miR-17-5p hsa-miR-18a-5p hsa-miR-20a-5p hsa-miR-20b-5p hsa-miR-93-5p hsa-miR-106b-5p	**miR-17/miR17-92**	**Acting on proteins associated with cell cycle regulation, proliferation, and stem cell renewal**	[37–41]
hsa-miR-15a-5p hsa-miR-15b-5p hsa-miR-16-5p hsa-miR-195-5p hsa-miR-21-5p hsa-miR-33a-5p	**miR-15**		

with a control. The experimental approach is to clone the 3′-UTR of the target gene of interest immediately downstream of the reporter gene. To study the direct interaction between the miRNAs and their 3′UTR putative sites we transfected HeLa cells with two commercially available plasmids containing the target 3′UTRs inserted downstream of the firefly luciferase reporter gene and renilla luciferase gene for normalization, and either one selected mimic miRNA, or scrambled miRNA negative control (Fig. 4b). Transfection efficiency was determined to be 100%.

4 Conclusions

In summary, we have developed a robust 3D in vitro hNSC differentiation assay which can be implemented to measure the gene expression of neural and glial markers, and quantify differential miRNA expression. The MiRNA profiles of hNSCs differentiated on 2D and 3D substrates showed similarity in term of miRNA types, but the degree and timing of miRNA regulation was significantly different for miRNAs involved in cell proliferation and

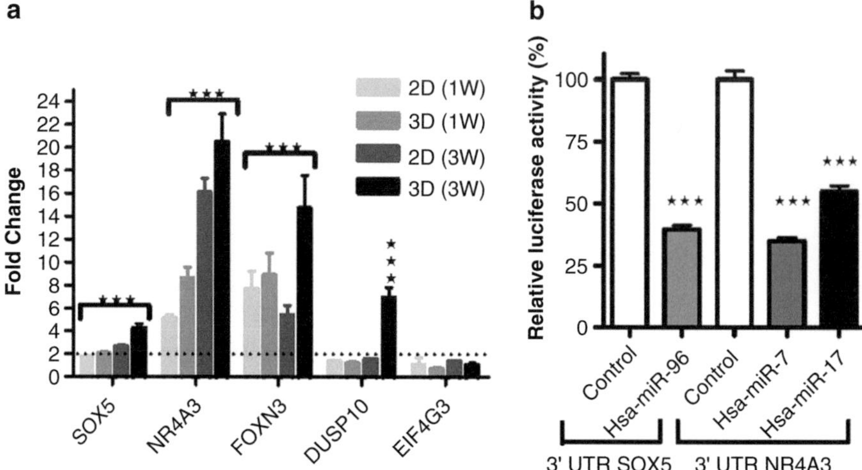

Fig. 4 Validation of miRNA target predicted genes (**a**) QRT-PCR analysis performed on 1W and 3W differentiated hNSCs cultured on 2D and 3D substrates and expressed as fold change compared with proliferative control. SOX5, NR4A3 (NOR1), and FOXN3 were found to be significantly up-regulated in differentiated samples compared with control proliferative cell cultures. In addition, DUSP10 was only found to be up-regulated in 3D differentiated cultures at 3W. Among the putative target mRNAs, only EIF4G3 was not up-regulated and therefore did not correlate with the prediction. Data are expressed as mean values $\pm$ SDMs and analysed by Student's *t*-test, * $p < 0.05$, ** $p < 0.01$, *** $p < 0.005$. (**b**) Dual luciferase report assay performed in HeLa cells transiently transfected with the SOX5 3′UTR, and NR4A3 3′UTR constructs and selected mimic miRNAs or miRNA control. The relative activity of firefly luciferase expression was normalized to renilla luciferase activity. Hsa-miR-96 caused a decrease in SOX5 3′UTR luciferase activity by 60.34% $\pm$4.79, and both hsa-miR-7 and hsa-miR-17 caused a decrease in NR4A3 3′UTR luciferase activity by 65.01% $\pm$ 4.07 and 45.11% $\pm$ 6.76, respectively, compared with the controls. Each bar represents values from three independent experiments, measured in triplicate. Data are expressed as mean values $\pm$ SDMs and analysed by Student's *t*-test, *** $p < 0.005$ [21]

maintenance of stemness. Using online available software analysis tools (DIANA Lab, PicTar, and TargetScans) a set of target mRNAs were identified and investigated. Two mRNAs were identified and validated as putative miRNA targets by qRT-PCR and 3′UTR dual luciferase reporter system. These observations suggest that 3D surface topography influences the molecular behaviour of hNSCs by modulating miRNAs associated with cell proliferation and stemness maintenance thus promoting hNSC differentiation.

Acknowledgement

This study was supported by ReNeuron (RENE.L). We acknowledge Julie Heward for helping in the preparation of hNSCs.

Table 4
Computational target gene predictions

		DIANALAB microT v3.0					PicTar	TargetScan
miRNA	Target gene	Rank	miTG score	Precision	SNR	Target sites/genes found	Score	Aggregate P_{CT}
hsa-miR-96	SOX5	2	64.73	1	7.3	1214/763	6.74	0.94
hsa-miR-183	DUSP10	1	26.44	0.72	1.62	302/190	4.75	0.85
hsa-miR-302a	NR4A3	27	19.86	0.84	6.22	863/473	3.05	NF
hsa-miR-182	FOXN3	3	37.4	0.94	6.87	1358/794	NF	0.96
hsa-miR-7	NR4A3	NF	NF	NF	NF	NF	1.61	0.17
hsa-miR-7	FOXN3	7	14.08	0.17	1.03	399/136	NF	0.88
hsa-miR-20a	NR4A3	9	26.13	0.85	9.44	1650/841	5.87	0.63
hsa-miR-20b	NR4A3	11	25.60	0.9	9	1955/973	5.87	0.63
hsa-miR-17	NR4A3	13	25.85	0.9	7.94	1928/961	5.38	0.63 + 0.37
hsa-miR-20a	EIF4G3	3	55.29	0.97	9.44	1650/841	NF	NF
hsa-miR-20b	EIF4G3	1	46.9	0.94	8	1955/973	NF	NF
hsa-miR-17	EIF4G3	1	46.87	0.94	7.94	1928/961	NF	NF

List of miRNA target prediction mRNAs. MiRNAs and target predictive genes were identified by algorithm analysis using Diana Lab, PicTar, and TargetScan. Scores (aggregate) are provided for each analysis

References

1. Ambros V (2004) The functions of animal microRNAs. Nature 431(7006):350–355. doi:10.1038/nature02871
2. Fire A, Xu S, Montgomery MK, Kostas SA, Driver SE, Mello CC (1998) Potent and specific genetic interference by double-stranded RNA in *Caenorhabditis elegans*. Nature 391 (6669):806–811. doi:10.1038/35888
3. Spivakov M, Fisher AG (2007) Epigenetic signatures of stem-cell identity. Nat Rev Genet 8 (4):263–271. doi:10.1038/nrg2046
4. Badylak SF (2002) The extracellular matrix as a scaffold for tissue reconstruction. Semin Cell Dev Biol 13(5):377–383
5. Buxboim A, Discher DE (2010) Stem cells feel the difference. Nat Methods 7(9):695–697. doi:10.1038/nmeth0910-695
6. Gelain F, Bottai D, Vescovi A, Zhang S (2006) Designer self-assembling peptide nanofiber scaffolds for adult mouse neural stem cell 3-dimensional cultures. PLoS One 1:e119. doi:10.1371/journal.pone.0000119

7. Li WJ, Tuli R, Okafor C, Derfoul A, Danielson KG, Hall DJ, Tuan RS (2005) A three-dimensional nanofibrous scaffold for cartilage tissue engineering using human mesenchymal stem cells. Biomaterials 26(6):599–609. doi:10.1016/j.biomaterials.2004.03.005
8. Lutolf MP, Gilbert PM, Blau HM (2009) Designing materials to direct stem-cell fate. Nature 462(7272):433–441. doi:10.1038/nature08602
9. Nur EKA, Ahmed I, Kamal J, Schindler M, Meiners S (2006) Three-dimensional nanofibrillar surfaces promote self-renewal in mouse embryonic stem cells. Stem Cells 24(2):426–433. doi:10.1634/stemcells.2005-0170
10. Ortinau S, Schmich J, Block S, Liedmann A, Jonas L, Weiss DG, Helm CA, Rolfs A, Frech MJ (2010) Effect of 3D-scaffold formation on differentiation and survival in human neural progenitor cells. Biomed Eng Online 9(1):70. doi:10.1186/1475-925X-9-70
11. Saha K, Pollock JF, Schaffer DV, Healy KE (2007) Designing synthetic materials to control stem cell phenotype. Curr Opin Chem Biol 11 (4):381–387. doi:10.1016/j.cbpa.2007.05.030
12. Hesse E, Hefferan TE, Tarara JE, Haasper C, Meller R, Krettek C, Lu L, Yaszemski MJ (2010) Collagen type I hydrogel allows migration, proliferation, and osteogenic differentiation of rat bone marrow stromal cells. J Biomed Mater Res A 94(2):442–449. doi:10.1002/jbm.a.32696
13. Budyanto L, Goh YQ, Ooi CP (2009) Fabrication of porous poly(L-lactide) (PLLA) scaffolds for tissue engineering using liquid-liquid phase separation and freeze extraction. J Mater Sci Mater Med 20(1):105–111. doi:10.1007/s10856-008-3545-8
14. Knight E, Murray B, Carnachan R, Przyborski S (2011) Alvetex(R): polystyrene scaffold technology for routine three dimensional cell culture. Methods Mol Biol 695:323–340. doi:10.1007/978-1-60761-984-0_20
15. Qutachi O, Vetsch JR, Gill D, Cox H, Scurr DJ, Hofmann S, Muller R, Quirk RA, Shakesheff KM, Rahman CV (2014) Injectable and porous PLGA microspheres that form highly porous scaffolds at body temperature. Acta Biomater 10(12):5090–5098. doi:10.1016/j.actbio.2014.08.015
16. Chan BP, Leong KW (2008) Scaffolding in tissue engineering: general approaches and tissue-specific considerations. Eur Spine J 17 (Suppl 4):467–479. doi:10.1007/s00586-008-0745-3
17. Bokhari M, Carnachan RJ, Przyborski SA, Cameron NR (2007) Emulsion- templated porous polymers as scaffolds for three dimensional cell culture: effect of synthesis parameters on scaffold formation and homogeneity. J Mater Chem 17:4088–4094
18. Bokhari M, Carnachan RJ, Cameron NR, Przyborski SA (2007) Novel cell culture device enabling three-dimensional cell growth and improved cell function. Biochem Biophys Res Commun 354(4):1095–1100. doi:10.1016/j.bbrc.2007.01.105
19. Krek A, Grun D, Poy MN, Wolf R, Rosenberg L, Epstein EJ, MacMenamin P, da Piedade I, Gunsalus KC, Stoffel M, Rajewsky N (2005) Combinatorial microRNA target predictions. Nat Genet 37(5):495–500. doi:10.1038/ng1536
20. Lewis BP, Burge CB, Bartel DP (2005) Conserved seed pairing, often flanked by adenosines, indicates that thousands of human genes are microRNA targets. Cell 120 (1):15–20. doi:10.1016/j.cell.2004.12.035
21. Stevanato L, Sinden JD (2014) The effects of microRNAs on human neural stem cell differentiation in two- and three-dimensional cultures. Stem Cell Res Ther 5(2):49. doi:10.1186/scrt437
22. Pollock K, Stroemer P, Patel S, Stevanato L, Hope A, Miljan E, Dong Z, Hodges H, Price J, Sinden JD (2006) A conditionally immortal clonal stem cell line from human cortical neuroepithelium for the treatment of ischemic stroke. Exp Neurol 199(1):143–155. doi:10.1016/j.expneurol.2005.12.011
23. http://diana.cslab.ece.ntua.gr/microT/
24. Maragkakis M, Alexiou P, Papadopoulos GL, Reczko M, Dalamagas T, Giannopoulos G, Goumas G, Koukis E, Kourtis K, Simossis VA, Sethupathy P, Vergoulis T, Koziris N, Sellis T, Tsanakas P, Hatzigeorgiou AG (2009) Accurate microRNA target prediction correlates with protein repression levels. BMC Bioinformatics 10:295. doi:10.1186/1471-2105-10-295
25. Chen K, Rajewsky N (2006) Natural selection on human microRNA binding sites inferred from SNP data. Nat Genet 38 (12):1452–1456. doi:10.1038/ng1910
26. http://pictar.mdc-berlin.de/
27. http://www.targetscan.org/
28. Bustin SA (2000) Absolute quantification of mRNA using real-time reverse transcription polymerase chain reaction assays. J Mol Endocrinol 25(2):169–193
29. Schmittgen TD, Livak KJ (2008) Analyzing real-time PCR data by the comparative C(T) method. Nat Protoc 3(6):1101–1108

30. Stevanato L, Hicks C, Sinden JD (2015) Differentiation of a human neural stem cell line on three dimensional cultures, analysis of MicroRNA and putative target genes. J Vis Exp 98. doi:10.3791/52410
31. Bentwich I (2005) Prediction and validation of microRNAs and their targets. FEBS Lett 579 (26):5904–5910. doi:10.1016/j.febslet.2005.09.040
32. Didiano D, Hobert O (2006) Perfect seed pairing is not a generally reliable predictor for miRNA-target interactions. Nat Struct Mol Biol 13(9):849–851. doi:10.1038/nsmb1138
33. Parsons XH, Parsons JF, Moore DA (2012) Genome-scale mapping of MicroRNA signatures in human embryonic stem cell neurogenesis. Mol Med Ther 1(2). doi: 10.4172/2324-8769.1000105
34. Wang Y, Keys DN, Au-Young JK, Chen C (2009) MicroRNAs in embryonic stem cells. J Cell Physiol 218(2):251–255. doi:10.1002/jcp.21607
35. Chen H, Shalom-Feuerstein R, Riley J, Zhang SD, Tucci P, Agostini M, Aberdam D, Knight RA, Genchi G, Nicotera P, Melino G, Vasa-Nicotera M (2010) miR-7 and miR-214 are specifically expressed during neuroblastoma differentiation, cortical development and embryonic stem cells differentiation, and control neurite outgrowth in vitro. Biochem Biophys Res Commun 394(4):921–927. doi:10.1016/j.bbrc.2010.03.076
36. Weeraratne SD, Amani V, Teider N, Pierre-Francois J, Winter D, Kye MJ, Sengupta S, Archer T, Remke M, Bai AH, Warren P, Pfister SM, Steen JA, Pomeroy SL, Cho YJ (2012) Pleiotropic effects of miR-183~96~182 converge to regulate cell survival, proliferation and migration in medulloblastoma. Acta Neuropathol 123(4):539–552. doi:10.1007/s00401-012-0969-5
37. Cioffi JA, Yue WY, Mendolia-Loffredo S, Hansen KR, Wackym PA, Hansen MR (2010) MicroRNA-21 overexpression contributes to vestibular schwannoma cell proliferation and survival. Otol Neurotol 31(9):1455–1462. doi:10.1097/MAO.0b013e3181f20655
38. Cirera-Salinas D, Pauta M, Allen RM, Salerno AG, Ramirez CM, Chamorro-Jorganes A, Wanschel AC, Lasuncion MA, Morales-Ruiz M, Suarez Y, Baldan A, Esplugues E, Fernandez-Hernando C (2012) Mir-33 regulates cell proliferation and cell cycle progression. Cell Cycle 11(5):922–933. doi:10.4161/cc.11.5.19421
39. Liu XS, Chopp M, Wang XL, Zhang L, Hozeska-Solgot A, Tang T, Kassis H, Zhang RL, Chen C, Xu J, Zhang ZG (2013) MicroRNA-17-92 cluster mediates the proliferation and survival of neural progenitor cells after stroke. J Biol Chem 288(18):12478–12488. doi:10.1074/jbc.M112.449025
40. Porrello ER, Johnson BA, Aurora AB, Simpson E, Nam YJ, Matkovich SJ, Dorn GW 2nd, van Rooij E, Olson EN (2011) MiR-15 family regulates postnatal mitotic arrest of cardiomyocytes. Circ Res 109(6):670–679. doi:10.1161/CIRCRESAHA.111.248880
41. Trompeter HI, Abbad H, Iwaniuk KM, Hafner M, Renwick N, Tuschl T, Schira J, Muller HW, Wernet P (2011) MicroRNAs MiR-17, MiR-20a, and MiR-106b act in concert to modulate E2F activity on cell cycle arrest during neuronal lineage differentiation of USSC. PLoS One 6 (1):e16138. doi:10.1371/journal.pone.0016138

Neuromethods (2017) 128: 59–71
DOI 10.1007/7657_2016_12

Published online: 27 October 2016

Study of miRNA Function in the Developing Axons of Mouse Cortical Neurons: Use of Compartmentalized Microfluidic Chambers and In Utero Electroporation

Patricia P. Garcez, Francois Guillemot, and Federico Dajas-Bailador

Abstract

The understanding of brain function requires the study of the cellular and molecular mechanisms that govern the growth and development of neuron connections. This is a process driven by the polarization of the neuron into a highly structured morphology with dendritic arborizations and a long axonal projection. In recent years, miRNAs have been described as key players in these processes, both at the level of axon development and synaptic plasticity. Here, we describe a microfluidic compartmentalized cortical neuron culture and the in utero electroporation techniques for the study of miRNAs and their role in axon development.

Keywords: Cortical neurons, Mouse cortex, Axon development, mRNA, miRNA, Local translation, Microfluidic chambers, In utero electroporation

1 Introduction

The specification and growth of an axon, with its guidance, branching, and synaptic terminal differentiation, are fundamental steps in the wiring of the approximately 10^{15} connections that are found in the brain [1, 2]. The highly polarized structure that is necessary for neuron connectivity makes them very sensitive to localized environmental cues and requires the dynamic segregation of proteins and lipids into the axonal and somato-dendritic compartments. For these reasons, it is now widely recognized that the localized translation of mRNAs has a crucial role in the regulation of neuron development and function [3]. It allows for gene expression to be spatially restricted within the somatic and axonal axis, and for that expression to have rapid temporal resolution [4]. In effect, the distal axon contains a diverse variety of mRNAs that can be regulated by various molecular mechanisms [4]. Among these, the role of miRNAs in the control of local translation and in neuron polarization and axon development has started to be unraveled [5–9]. This is particularly relevant since developmentally regulated miRNAs in the cortex appear to be evolving far more rapidly than other classes of genes, and might account for a significant part of the expansion in cognitive capacity in humans [6].

The ability to regulate the expression, localization, and function of miRNAs by neuronal signaling processes offers them a unique opportunity to serve as agents of developmental and adaptive change [6, 7]. Experimentally, although multiple studies have addressed the role of miRNAs in models of activity-induced synapse plasticity [10], the function of miRNAs in axon development has been comparatively less investigated. This is partly due to the greater difficulty of addressing localized axonal processes in experimental conditions. For this reason, novel culture approaches have been devised to investigate some of the specific physiological processes observed in neuronal function. In particular, compartmentalized microfluidic devices have become an increasingly useful tool in neuronal cell biology due to the ability to precisely control and monitor cellular microenvironments [11, 12]. Cortical neurons can thus be cultured for long periods in conditions that allow the manipulation of the axon environment independent of the cell body. In the last few years, we have used the microfluidic devices first described by the lab of Noo Li Jeon [11, 13] and adapted their protocols to study the localized effects of miRNAs in the axons of developing cortical neurons.

Although the use of advanced culture devices permits the elucidation of detailed molecular mechanisms in the function of miRNAs, it is also important to assess these findings using whole brain in vivo approaches. In this context, the use of in utero electroporation allows the examination of axonal growth in the intricate and complex context of whole brain development. This technique permits cell tracking by expression of reporter genes, such as eGFP, together with the manipulation of gene expression using overexpression constructs, siRNA, or microRNAs [7, 14, 15]. The in utero electroporation technique has remarkable advantages when compared with transgenic mice. First, it allows the acute regulation of gene expression, while avoiding the genetic compensation that might occur in the process of transgenic animal generation. Second, it is a very fast and reproducible procedure, amenable to candidate genes screening approaches. Third, it is a powerful tool to restrict spatial and temporal gene function analysis, which can also be used in association with transgenic floxed mouse to knock-out genes with Cre construct electroporation [16].

In the last decade, in utero electroporation has been extensively adapted to study multiple developmental processes occurring in different regions of the brain, including cortical progenitors proliferation, neuronal polarization, and neuronal migration [17, 18]. Late developmental aspects of neuronal maturation such as axonal guidance, growth, and dendritic arborization have also been studied using this technique [7, 19–21], thus addressing some key regulatory processes targeted by miRNAs.

In this chapter, we aim to describe in detail how microfluidic culture models and in utero electroporation can be used as complementary approaches to investigate the function of miRNAs in the development of mouse cortical axons.

2 Materials

2.1 Cortical Neurons Primary Cultures

Poly-L-Lysine or Poly-L-ornithine (cat. No: P1274 from Sigma). Neurobasal media, GlutaMAX and B27 supplement from Invitrogen.

Fine forceps, extra fine and spring scissors, and curved Iris spatula from Fine Science Tools (F.S.T.).

2.1.1 Compartmentalized Microfluidic Devices

Microfluidic device (cat. No. SND150, Xona Microfluidics) and glass coverslips of similar dimension.

2.1.2 miRNA Inhibitors and Neuron Transfection

Endotoxin free pMAX-GFP construct from Amaxa (for transfection of 1.5 μg per well using Lipofectamine 2000, 5 μl in 2 ml of final volume). PNA-miRNA inhibitors with cell penetrating peptide from Panagene.

2.2 In Utero Electroporation

2.2.1 DNA Injection

Endotoxin-free pCIG-GFP construct (1 μg/μl), locked nucleic acids (LNA) non-targeting control, or experimental miRNA inhibitors (50 nM) mixed with Fast green water solution (0.05 %, Sigma).

Microloader tips (Eppendorf).

FemtoJet express microinjector (Eppendorf).

Glass capillary needles (1 mm O.D. × 0.58 mm I.D.; Harvard Apparatus).

2.2.2 Surgery

Analgesic (Carprophene—5 mg/kg subcutaneously), plus anesthetic induction chamber and mask.

Isoflurane (Anesthesia) and disinfectant (Chlorhexidine).

Electric Razor, Heating Pads (2), sterile drapes, cotton buds, gloves, mask, and swabs.

Surgical instruments: Scalpel, needle holder with suture cutter forcep, Graefe forceps, and Ring forceps (F.S.T.). Vicryl absorbable suture 6/0 (Ethicon Inc.).

2.2.3 Electroporation

BTX single waveform Electroporation System (Harvard Apparatus). Capillary holder (Eppendorf) and electrodes (Harvard Apparatus).

2.2.4 Tissue Processing

4 % PFA in PBS.

Fluorescent binocular.

Vibratome and vibratome blades (Leica).

Slides, coverslips, and Mounting medium (Aqua Polymount; Polysciences Inc).

2.2.5 Immunofluorescence Studies

Anti-GFP antibody (R&D; catalog number: AF4240).

Donkey Anti-Sheep IgG Secondary Antibody, Alexa Fluor 488 conjugate (Life technology, catalog number: A-11015)

DAPI (Invitrogen).

3 Methods

3.1 Part (1): Study of miRNA Function in Mouse Cortical Neurons Grown in Compartmentalized Microfluidic Devices

3.1.1 Microfluidic Device Preparation and Assembly (2 Days)

- Glass coverslips are sterilized by submerging in 70 % ethanol and then left to dry inside a laminar flow tissue culture cabinet (TC cabinet).
- Once dried, coverslips are placed in culture dishes and covered with poly-L-Lysine or poly-L-ornithine solution (overnight incubation at 37 °C, or for a minimum of 3–4 h).
- After coating, coverslips are washed twice with sterile H_2O and left to dry completely inside the TC cabinet.
- Before assembly, new microfluidic chambers are left overnight in Neurobasal media (*see* **Note 1**), then washed in water, sterilized in 70 % ethanol, and left to dry inside the TC cabinet (this last step is done in parallel with the coverslip sterilization described above). Autoclaving is not recommended as it could affect bonding between PDMS and glass surface.
- For the assembly, fully dry devices are placed on the coated glass coverslip, allowing a perfect seal to form between the PDMS and the glass surface (*see* **Note 2**). This type of seal is reversible. Culture media is immediately applied to the channels (by gently pipetting at an angle into them) and in the wells of the chambers (*see* Fig. 1 for more details).
- The assembled microfluidic device is left overnight (or for at least 3–4 h) to allow uniform flow of media inside the microgrooves. For this, a slight differential volume between wells is established (see below and Fig. 1).

3.1.2 Cell Culture and Chamber Seeding (1 Day)

- Cortical neurons are obtained from embryonic day (E) 16–18 mice by making a neuron suspension after the dissection of four cortices (from two brains). The cortical tissue is dissociated in 3 ml of Neurobasal media to obtain a final neuron density of approximately 1.5×10^6 cells/ml.
- Before seeding neurons inside the microfluidics device, the media left overnight has to be removed from wells, avoiding the aspiration from inside the channels. Immediately add 20 μl of neuron suspension to one side of the soma channels, and repeat action for the other side of the same channel.
- Place the device into a humidified 37 °C incubator for 20–30 min to permit neurons to attach to the coated surface inside the device channel.

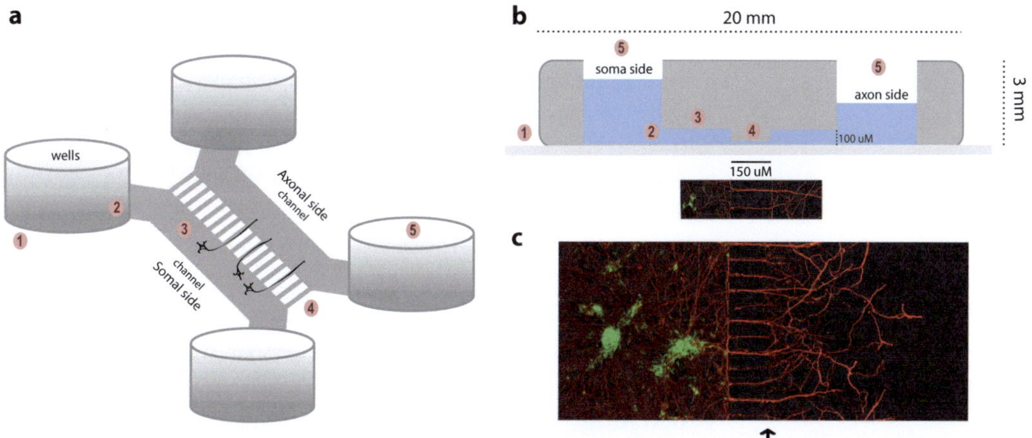

Fig. 1 (**a** and **b**) Schematic illustrations of a microfluidic culture chamber made of a PDMS mold with a pattern of channels (axonal v somal) connected by microgrooves and placed on a glass coverslip (*point 1*). Two wells on each side feed the channels and allow for a differential volume (*point 5*) to be established [11]. This generates a sustained flow between compartments that counteracts diffusion and establishes fluidic isolation between axonal and somal sides. Neurons are seeded by gently pipetting into the channel on the soma side (*point 2*), thus allowing cell bodies to attach inside the channel (*point 3*) and start their morphological differentiation into polarized neurons. Some of the axons grow into the microgrooves (*point 4*) and extend into the axonal channel on the opposite side of the device. (**c**) The image shows an immunofluorescent staining of cortical neurons cultured in microfluidic devices. Acetylated tubulin (*red*) preferentially labels axons, while the transcription factor Otx-1 (*green*) localizes in the cell bodies of cortical neurons. *Black arrow* indicates the point of exit from microgrooves. Axonal growth is estimated by measuring individual axons from this point up to the growth cone

- Use a microscope to check neuron attachment to the surface inside the channel before adding a further 60 μl of neuron suspension to each well (not directly into the channel). This neuron population will provide trophic support for neurons growing inside the somal channel. After this, add a further 50 μl of fresh media to somal wells and 110 μl to each of the wells of future axon side.
- In order to achieve a fluidically isolated microenvironment in the axon/cell body compartments a difference in volume between sides must be established. Hydrostatic pressure differences create a continuous flow across microgrooves preventing diffusion in the opposite direction. According to Park and collaborators [13] a 30 μl difference in volume will prevent diffusion of molecules larger than 1 kDa, while 50 μl difference should be used for smaller molecules.
- Media levels should be checked every 24 h and replenished accordingly (*see* **Notes 3** and **4**).
- Maintenance of fluidic isolation allows the addition of drugs/inhibitors to one side of the chambers to measure local effects in the neurons (see below).

3.1.3 Axon Specific Stimulation/Transfection

- The maintenance of fluidic isolation provides an experimental setup for the manipulation of local processes that occur in the axon compartment of the neuron. Importantly, the RNAi pathway has been shown to be functional in developing axons, thus allowing the selective knockdown of mRNAs after specific siRNA transfection in the axon compartment [22–24]. Axonal transfection can also be performed with miRNA mimics or inhibitors made using modified stable nucleic acids, i.e., locked nucleic acids (LNAs), peptide nucleic acids (PNAs).
- In addition to standard transfection protocols with Lipofectamine 2000, local loss of function of axonal miRNAs can also be achieved using miRNA inhibitors conjugated to cell penetrating peptides [7]. These conjugated oligonucleotides avoid the need for axonal transfection and can be added directly to the axonal channel of a microfluidic device.
- Although fluidic isolation between axonal and somal compartments can be easily achieved by differential hydrostatic pressure, it is important to consider that the axon itself offers a potential source of communication between the growth cone and cell bodies after transfection in the axon compartment. Importantly, Hengst and coauthors [22] demonstrated that FITC-labeled siRNA transfected in the axon compartment appeared to be incorporated into relatively immobile punctate structures within axons, without significant diffusion. As a result, knock-down effects were only observed in the axonal side of the chambers [22, 24]. Although this would indicate the lack of retrograde transport for transfected siRNA, it is important to test this in each experimental setup, or to perform functional studies where cell body and axon-specific transfections are compared [7].
- The axonal effects of neurotransmitters, receptor agonist/antagonist, and/or growth factors could be studied after local application into the axon channel. In this way, axonal mechanisms can be studied independently of primary cell body influence [7, 25]. Crucially, one has to consider the time course of stimulation to discard effects where axonal signals are transmitted to cell body and a subsequent outcome is observed in the axon.

3.1.4 Immunofluorescence Studies

- The neuronal cell bodies and their axons can be fixed either before or after removing the microfluidic device. In both cases, maximum care needs to be taken when detaching the device, as this could damage the extending axons.
- Before fixing, wash with PBS and then add freshly made solution of 4 % paraformaldehyde in PBS, 30 min at room temperature (RT) (*see* **Note 5**).

- Wash with PBS three times and incubate with PBS/0.2 % Triton for 30 min to permeabilize cells.
- Wash with PBS/0.1 % Triton three times before adding blocking solution (3 % BSA, 30 min at RT).
- After blocking, incubate primary antibodies overnight at 4 °C in a humidified chamber.
- Wash three times with PBS/0.1 % Triton and incubate with secondary antibody for 1 h at RT.
- Wash and mount before analysis of axon length or protein levels (see below).

3.1.5 Microfluidic Data Analysis

- The use of immunofluorescence techniques allows the detection of changes in the levels of proteins, mRNAs, or miRNAs, either at the axon or cell body compartments. For example, local and rapid axonal stimulation with BDNF or glutamate can decrease levels of miR-9 and JIP-1 protein respectively [7, 25].
- In addition, axon growth can be estimated by measuring the axon length from the microgrooves exit into the axonal channel, thus providing a functional assessment of axon development (*see* Fig. 1).
- The microfluidic device also constitutes a great experimental model for the study of directional axonal transport [26], and the functional implications that it can have in the regulation of local translation by axonal miRNAs.

3.1.6 RNA Harvest

The microfluidic-based compartmentalized chambers can be used to selectively isolate RNA material from the axons of cortical neurons, thus allowing the performance of RT-PCR and/or next-generation sequencing on a specific population of neuronal RNAs. The protocol for axonal RNA isolation using Trizol is as follows:

- Remove media from somal and axonal channels and wash the latter with PBS.
- Add 100 μl of Trizol into each well of axonal side and leave to incubate for 3–5′ (make sure Trizol enters into the axonal channel of the device).
- Collect Trizol from both wells and the channel of axonal side and vortex vigorously (15″).
- Add 1/5 volume of chloroform (invert a couple of times) and spin down at max speed (4 °C 16′).
- Collect aqueous upper phase, and add equal volume of chloroform before vortexing for 15″.

- Spin down at max speed, 4 °C 10′, before collecting aqueous phase and adding equal volume of isopropanol (keep at −20 o/n).
- Next day spin down at max speed, 4 °C 30′, remove and discard supernatant.
- Add 200 μl of 70 % ethanol (in RNAse free water).
- Spin down at max speed, 4 °C 10′.
- Repeat ethanol wash.
- Dry pellet and resuspend in RNAse free water (~10 μl).

3.2 Part (2): Mouse In Utero Electroporation for the Study of Cortical Axon Development In Vivo

3.2.1 Surgery Preparation

- Prepare a clean after surgery cage and place a heat pad (37 °C) to keep females from same experimental groups.
- Prepare surgery theater area by disinfecting the surface and covering with sterile drapes. Put syringe for sterile PBS (37 °C) and sterile surgical instruments on a sterile drape, while electrodes are placed into a PBS beaker.
- Load the DNA solution into a glass pipette needle with a micro loader tip. Connect the needle with the capillary holder.

3.2.2 Preoperation Area Preparation (Non-sterile)

- Anesthetize the pregnant mouse with isoflurane in oxygen (2 l/min) using an induction chamber.
- Move the anesthetized animal to a presurgery mask and give Carprieve (Carprofen) subcutaneously (0.05 % PBS solution; 10 μl/g). Remove the hair of the animal 2 cm around the incision site with the electric razor. Disinfect the skin with chlorhexidine using a new swab each time and repeat three times. Apply eye drops to avoid eye dehydration.

3.2.3 Surgery

- Transfer the mouse to the surgery area mask on the heating pad.
- Assess anesthesia by pedal withdrawal reflex loss.
- Change gloves to sterile ones.
- Isolate the belly area by using a sterile drape with a small hole over the abdomen.
- With a sterile scalpel, make vertical incision along the midline (~1 in.). Using another scalpel or scissors, make an incision along the *linea alba* in the abdominal muscle (white line). In case of bleeding, use sterile swab with gentle pressure.
- Using ring forceps, carefully place the embryos out of the abdominal cavity on the top of the sterile drape.
- Use the pre-warmed PBS to keep the embryos hydrated at all times.

3.2.4 Injection and Electroporation

- Apply PBS (37 °C) before and after the electroporation of each embryo.
- Using ring forceps, carefully bring the embryo closer to the uterine wall and inject approximately 1 μl in the ventricle close to the midline of one hemisphere.
- Position the electrodes around the sides of the brain, with the positive electrode directed to the injected ventricle/hemisphere (*see* **Note 6** and Fig. 2a).
- Apply the electrical pulses: 5× 30 V; 50 ms; with 1 s interval for cortical electroporation E14.5 (*see* **Note 7**).
- Place the electrode back in the PBS.
- Repeat process for each of the embryos.

3.2.5 End of the Surgery

- Carefully place the uterine horns back into the abdominal cavity before performing continuous abdominal muscle suture, followed by skin discontinuous suture (*see* **Note 8**).
- Place the animal in the recovery warm chamber until complete recovery from anesthesia and then in cage on the heating pad.

3.2.6 Post-surgery Care

- Animals should recover within 10 min and should behave normally without signs of pain or distress.
- Monitor animals for 24–48 h after surgery. Pain or distress should be assessed by animal behavior (*see* **Note 9**).

3.2.7 Tissue Processing

- For axonal growth studies embryos should be removed 4 days after electroporation.
- Select the electroporated brains by GFP visualization under a fluorescent binocular microscope.
- Dissect and fix the brains overnight in 4 % PFA.
- Section the brain coronally using a vibratome (50 μm sections).
- Use GFP antibody detection to enhance the GFP signal. For the immunocytochemistry block the sections with 10 % donkey serum diluted in PBS 0.1 % Triton for 30 min and follow by overnight incubation anti-GFP diluted in PBS 0.1 % triton (1:700) at 4 °C.
- After primary antibody incubation, wash with PBS 0.1 % Triton (3 × 10 min at RT) and incubate with secondary antibody (1:500) and DAPI (1:10,000) at RT for 2 h. Wash the sections with PBS (3 × 10 min at RT) (*see* **Note 10**).
- Mount in Aqua PolyMount (*see* **Note 11**).

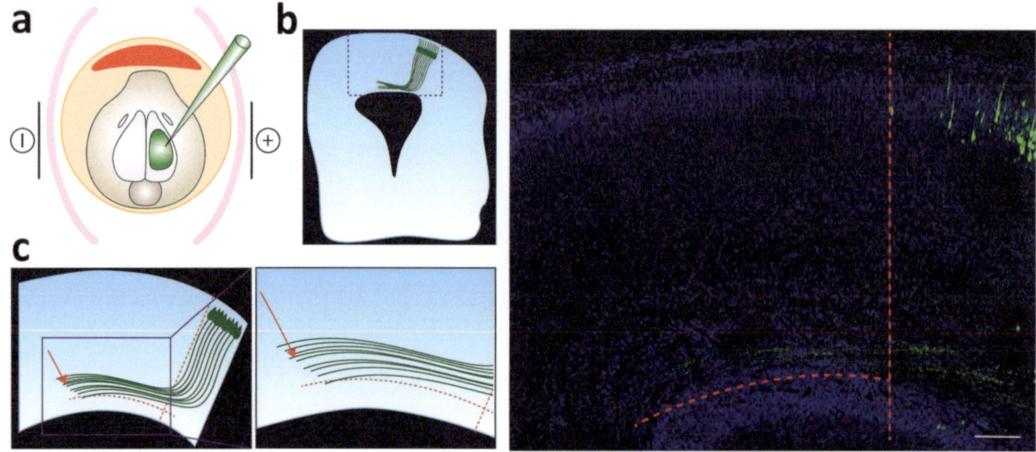

Fig. 2 In utero electroporation for axon growth studies in vivo. (**a**) Schematic representation of the electrode position during in utero electroporation. First, DNA is injected on the lateral ventricle in one of the cortical hemispheres, close to the midline. Second, electroporation is performed by positioning the electrodes around and parallel to the head of the embryo, avoiding the placenta (represented here in *red*), and applying the five electrical pulses of 30 V during 50 ms with 1 s interval between the pulses. (**b**) Schematic representation of the electroporated cortical area in a coronal hemi section. (**c**) Schematic diagram and high magnification image (*right panel*) of the electroporated GFP+ neurons and their axons growing in the intermediate zone medially toward the midline. *Arrow* points to the tip of the growing axons. The *vertical dashed line* represents the most medial electroporated cell, which is taken as a reference start point for the measurement of growing axons. Each GFP positive axon is measured from the reference point, with the *horizontal dashed line* representing the axons measured from the *vertical line* to the tip of the growth cone (*arrow*)

3.2.8 Data Analysis

- Analyze the data from at least three independent experiments for each condition.
- Use a confocal microscope to acquire images of the electroporated hemisphere (20× objective, 100 nm/pixel, 1 μm optical slice.
- Use ImageJ to measure the axonal length in the intermediate zone of at least 200 axons in each condition (*see* **Note 12** and Fig. 2).

4 Notes

4.1 Part (1): Microfluidic Neuronal Cultures

1. Before starting with the assembly protocol it is very important to immerse chambers overnight in Neurobasal media to remove any trace amounts of potentially toxic chemicals left on the PDMS surface.
2. Both the coverslips and chambers need to be completely dry to guarantee a perfect seal. Washing and drying steps need to be carried out in a TC cabinet to avoid contamination and minimize dust/debris adhering to PDMS and glass. Careful

manipulation is needed not to damage the microgrooves and the bottom side of the chambers during handling.

3. Excessive evaporation of media can rapidly affect viability of neurons in devices and induce osmotic imbalances. For this reason, it is important to use a well-humidified cell incubator and to keep three to four assembled chambers inside one 10 cm Petri dish, thus minimizing media loss. Alternatively 1–2 ml of water can be added to the Petri dish, preferably avoiding contact with the chambers.
4. In order to improve growth and survival of cortical axons, conditioned media (from similarly staged cortical neurons grown simultaneously on normal culture wells) could be added to the axon side of the microfluidic device.
5. Freshly made paraformaldehyde solution is needed for fixing axons. Paraformaldehyde made from frozen aliquots did not provide the same level of tissue fixation, particularly in the case of axonal structures.

4.2 Part (2): In Utero Electroporation

6. In order to increase survival of embryos, care should be taken during manipulation, avoiding bleeding due to forcep pulling. The electrodes should not be positioned across the placenta, and the two embryos located medially and closer to the cervix (one in each uterine horn) should not be electroporated, as manipulation of this area may cause abortion.

   To facilitate the experimental analysis it is important to electroporate the same cortical area in each brain. The somatosensory cortex, which is located dorsolaterally, is a relatively easy area to target by positioning the electrodes parallel to the head, as is represented in Fig. 2a, b.
7. There are advantages of using E14.5 developmental stage. It is an embryonic age when only upper granular layer neurons are born, which make cortico-cortical projection neurons. As a result, their axons extend only toward the mid line to form the corpus callosum, simplifying the axonal growth measurements. In addition, the developmental age at which the upper granular layer neurons are born is an ideal stage to perform in utero surgery. Finally, the uterus wall is more transparent compared to earlier stages, facilitating the midline visualization and lateral ventricle injection.
8. Ideally, the whole surgical procedure should take no longer than 30 min. In case the stiches come out in the days following the experiment, it is advisable to re-stich the animal. If there is a scar, it is necessary to remove any scaring tissue from the wound edges with a sterile scalpel as otherwise the re-stitching will not be effective. The analgesic dose should be repeated

(Carprofen 5 mg/kg, subcutaneously) in case of pain or animal distress signs 24 h after surgery.

9. Animals with persisting signs of pain or distress should be culled.
10. All the immunohistochemistry steps are done with free floating sections using gentle agitation.
11. The slide with sections should be dry before adding the Aqua Poly Mount solution and the coverslips. Bubbles should be avoided by gentle manipulation of the coverslip.
12. ImageJ is free software that can be downloaded at imagej.nih.gov. The most medial electroporated cell was taken as a reference point to start measuring the axons (*see* Fig. 2b). Axons were identified using the z stack, labeled with the line tool in Image J and the result length was identified with the "measure" function.

Acknowledgments

We thank Dr. Yolanda Saavedra Torres for the comments on the manuscript and Wai Han Wong for technical support.

References

1. Williams RW, Herrup K (1988) The control of neuron number. Annu Rev Neurosci 11:423–453
2. Lewis TL, Courchet J, Polleux F (2013) Cell biology in neuroscience: cellular and molecular mechanisms underlying axon formation, growth, and branching. J Cell Biol 202: 837–848
3. Martin KC, Ephrussi A (2009) mRNA localization: gene expression in the spatial dimension. Cell 136:719–730
4. Jung H, Yoon BC, Holt CE (2012) Axonal mRNA localization and local protein synthesis in nervous system assembly, maintenance and repair. Nat Rev Neurosci 13: 308–324
5. Fabian MR, Sonenberg N (2012) The mechanics of miRNA-mediated gene silencing: a look under the hood of miRISC. Nat Struct Mol Biol 19:586–593
6. McNeill E, Van Vactor D (2012) MicroRNAs shape the neuronal landscape. Neuron 75:363–379
7. Dajas-Bailador F et al (2012) microRNA-9 regulates axon extension and branching by targeting Map1b in mouse cortical neurons. Nat Neurosci 15:697–699
8. Kaplan BB, Kar AN, Gioio AE, Aschrafi A (2013) MicroRNAs in the axon and presynaptic nerve terminal. Front Cell Neurosci 7:126
9. Chiu H, Alqadah A, Chang C (2014) The role of microRNAs in regulating neuronal connectivity. Front Cell Neurosci 7:283
10. Olde Loohuis NFM et al (2011) MicroRNA networks direct neuronal development and plasticity. Cell Mol Life Sci 69:89–102
11. Taylor AM et al (2005) A microfluidic culture platform for CNS axonal injury, regeneration and transport. Nat Methods 2:599–605
12. Millet LJ, Gillette MU (2012) New perspectives on neuronal development via microfluidic environments. Trends Neurosci 35:752–761
13. Park JW, Vahidi B, Taylor AM, Rhee SW, Jeon NL (2006) Microfluidic culture platform for neuroscience research. Nat Protoc 1:2128–2136
14. Saito T, Nakatsuji N (2001) Efficient gene transfer into the embryonic mouse brain using in vivo electroporation. Dev Biol 240:237–246
15. Tabata H, Nakajima K (2001) Efficient in utero gene transfer system to the developing mouse brain using electroporation: visualization of neuronal migration in the developing cortex. Neuroscience 103:865–872

16. Matsuda T, Cepko CL (2007) Controlled expression of transgenes introduced by in vivo electroporation. Proc Natl Acad Sci U S A 104:1027–1032
17. Azzarelli R et al (2014) An antagonistic interaction between PlexinB2 and Rnd3 controls RhoA activity and cortical neuron migration. Nat Commun 5:3405
18. Garcez PP et al (2015) Cenpj/CPAP regulates progenitor divisions and neuronal migration in the cerebral cortex downstream of Ascl1. Nat Commun 6:6474
19. Cancedda L, Fiumelli H, Chen K, Poo M-M (2007) Excitatory GABA action is essential for morphological maturation of cortical neurons in vivo. J Neurosci 27:5224–5235
20. Wang C-L et al (2007) Activity-dependent development of callosal projections in the somatosensory cortex. J Neurosci 27:11334–11342
21. Haas MA et al (2013) Alterations to dendritic spine morphology, but not dendrite patterning, of cortical projection neurons in Tc1 and Ts1Rhr mouse models of down syndrome. PLoS One 8:e78561
22. Hengst U, Cox LJ, Macosko EZ, Jaffrey SR (2006) Functional and selective RNA interference in developing axons and growth cones. J Neurosci 26:5727–5732
23. Hengst U, Deglincerti A, Kim HJ, Jeon NL, Jaffrey SR (2009) Axonal elongation triggered by stimulus-induced local translation of a polarity complex protein. Nat Cell Biol 11:1024–1030
24. Gracias NG, Shirkey-Son NJ, Hengst U (2014) Local translation of TC10 is required for membrane expansion during axon outgrowth. Nat Commun 5:3506
25. Dajas-Bailador F, Bantounas I, Jones EV, Whitmarsh AJ (2014) Regulation of axon growth by the JIP1-AKT axis. J Cell Sci 127:230–239
26. Zala D et al (2013) Vesicular glycolysis provides on-board energy for fast axonal transport. Cell 152:479–491

Neuromethods (2017) 128: 73–88
DOI 10.1007/7657_2016_13

Published online: 28 October 2016

Functional Analysis of Cortical Neuron Migration Using miRNA Silencing

Pierre-Paul Prévot, Marie-Laure Volvert, Alexander Deiters, and Laurent Nguyen

Abstract

MicroRNAs (miRNAs) are endogenous, single-stranded ~21-nucleotide-long noncoding RNAs that have emerged as key fine-tuning posttranscriptional regulators of gene expression. The validation of miRNA-target interactions in animal model systems is not trivial, especially in the developing cerebral cortex. Induction of miRNAs loss-of-function is the ideal way to study their physiological role in vivo. Although it has been accepted that the dramatic brain phenotype of the Dicer conditional knockout mouse resulted from loss of mature miRNAs, functional connections to individual miRNAs need to be carried out. In this chapter, we compare three methods that are currently used to promote the loss-of-function of selected miRNAs in the developing cerebral cortex: genetic knockouts, small molecule inhibitors, and miRNA sponges. As an example, we are presenting some data obtained with different miRNA-loss of function approaches that support a role for miR-22 and miR-124 in radial migration and multipolar–bipolar transition of cortical projection neurons. These distinct loss of function methods provide complementary information and results indicate that, depending of the scientific question, one can choose between these methods to analyze the role of selected miRNAs in cortical development.

Keywords: miRNA, Loss-of-function, Corticogenesis, Electroporation, Small molecule inhibitor, Sponge, Dicer

1 Introduction

MicroRNAs (miRNAs) are endogenous, single-stranded ~21-nucleotide-long noncoding RNAs that have emerged as key post-transcriptional regulators of gene expression [1, 2]. MiRNA biogenesis starts with transcription of a primary miRNAs (pri-mRNA) by RNA polymerase II. Pri-miRNAs are then processed by Drosha/DGCR8 complexes. The resulting precursor hairpin of 70 nucleotides, the pre-miRNA, is exported from the nucleus by Exportin-5–Ran-GTP [3], and processed into a mature small duplex miRNA by the RNase III enzyme Dicer1 [4], in complex with the double-stranded RNA-binding protein TRBP (HIV-1 TAR RNA binding protein). The functional strand of the mature miRNA is loaded, together with Argonaute (Ago2) proteins, into the RNA-induced silencing complex (RISC). There, on the basis of sequence complementary, it guides RISC to target mRNAs, in order to inhibit their expression through mRNA cleavage, translational repression, or deadenylation [2, 5, 6].

By doing so, miRNAs fine-tune gene expression levels by repressing translation of their mRNA targets. In mammals, miRNAs are predicted to control the activity of more than 60 % of all protein-coding genes [7, 8]. Consequently, miRNA have emerged as novel posttranscriptional regulators of gene expression. They are involved in almost every biological process investigated to date, including embryonic development and pathological conditions such as cancer [2, 9]. Given its complex architecture, it is not surprising that miRNAs are abundantly expressed in the brain, where they have been found to play important roles in the regulation of neurogenesis, including the coordination of the successive steps of corticogenesis [10–13].

Corticogenesis begins at embryonic day (E) 11 in mouse cortices at the ventricular zones of the dorsal telencephalon and ends approximately at E18. The cerebral cortex develops from the dorsal telencephalon, which comprises several types of progenitors located around the lateral ventricles. At the onset of corticogenesis, neuroepithelial cells (NE) convert into radial glial cells (collectively named apical progenitors or APs) [14]. Initially, most APs undergo proliferative division to expand their pool but they later start to divide asymmetrically (differentiative division) to self-renew and give birth to neurons, either directly or indirectly through generation of basal progenitors (BPs) [14]. BP settle in the subventricular zone (SVZ) and divide symmetrically to generate either two post-mitotic neurons that migrate along radial glia fibers to reach the cortical plate (CP) or, more rarely, additional pairs of progenitor cells [15, 16]. At the end of corticogenesis most divisions are differentiative divisions in order to produce neurons. Noctor and colleagues demonstrated that multipolar–bipolar conversion of projection neurons is further required for the appropriate glia-guided locomotion to settle in the cortical plate and lead to a six cortical layers in an inside-out pattern [17].

Studies of miRNA function through targeting can be conducted in a variety of ways, including genetic deletion. For instance, mouse lines carrying a conditional deletion of Dicer in the telencephalon have been established and revealed critical roles for mature miRNAs in cortical neurogenesis [18–21]. However, the deletion of Dicer and the subsequent lack of most mature miRNAs lead to dramatic phenotypes and a major challenge is now to analyze the individual function of hundreds of miRNAs. One solution is to profile miRNAs involved in development. Analysis of miRNA expression patterns—for example, though traditional microarray technology—can be used to elucidate the roles of specific miRNA(s) in the regulation of their target(s) [22–27]. Similarly to mRNAs, expression of miRNAs varies widely between cells, tissues, organs and species during development [2]. Several miRNAs are abundant in the developing cerebral cortex and some show a dynamic expression that correlates with important developmental milestones of the cortex [28, 29].

Although it has been generally accepted that the phenotype of conditional knockout (cKO) Dicer mice resulted from loss of mature miRNAs, functional connections to individual miRNAs need to be carried out. The regulatory roles of miRNAs are reflected in the cellular functions of the miRNA target genes. Several laboratories have been developing bioinformatics approaches [30–33] but the number of putative targets for miRNA-mediated silencing so far outnumbers those that have been experimentally confirmed. Therefore, acute modulation of single miRNA activity needs to be performed in vitro and in vivo to assign specific functions and targets to these molecules.

The validation of miRNA–target interactions in animal model systems is not trivial. Inducing a loss of function of miRNAs is a powerful functional genomic approach and the best way to identify the biological implications of physiological miRNA levels. Generation of KO mice for individual miRNA is labor intensive despite the recent development of CRISPR/Cas9 technology [34, 35] and has not been always proved to be successful [2, 36]. MiRNAs are indeed generally encoded by multigene family members and the loss of function of one miRNA member is often compensated by redundant functions of other miRNAs. Alternative methods and strategies have been developed to block the regulatory functions of all miRNA members of a given family, including chemically modified miRNA inhibitors (antisense miRNA oligonucleotides—AMOs) or miRNA sponges. These molecules have recently proven to be effective in blocking functions of specific miRNA families, and have been deployed in cortical development analyses through in utero electroporation and subsequent ex vivo brain slice experiments [37–39].

Chemically modified oligonucleotides have been widely used in loss-of-function studies of miRNAs. Based on an antisense strategy, AMOs complementary to the miRNAs act as competitive inhibitors to endogenous target mRNA binding sites, leading to a suppression of miRNA functions. AMOs have been developed and demonstrated to be very specific and potent inhibitors of targeted miRNAs [40]. The major modification in AMOs is 2′-*O*-methylation (2′-OMe) of the ribose, sometimes combined with other types of modifications such as phosphorothioate linkages near 5′ and 3′ ends in order to stabilize the antisense molecules. Deiters and colleagues have recently developed light-activated antagomirs, by adding a light-removable nucleobase-caging group through chemical introduction of a 2′-OMe NPOM (6-nitropiperonyloxymethyl)-caged uridine phosphoramidite into the AMO [41].

These reagents enable photochemical regulation of miRNA function with high spatial and temporal resolution.

The principle of 'miRNA sponges' has been developed by Ebert and colleagues [42, 43]. MiRNA sponges are transcripts expressed from strong promoters, containing multiple tandem complementary binding sites to a selected miRNA, but with a bulge or mismatch in the seed site. Unlike synthetic antisense molecules, miRNA sponges with bulges theoretically prevent the degradation of the RNA sponge and efficicently sequester the miRNA, thereby releasing the translational repression of mRNA targets.

In this chapter, we compare three loss-of-function experiments, which lead to the conclusion that miR-22 and miR-124 are required to ensure migration of cortical projection neurons.

2 Material and Methods

2.1 Sponge Construction

Cluster sponge elements were generated by annealing multiple oligonucleotides to obtain a sequence containing six copies of a specific miRNA binding site (MBS) for miR-22, miR-124, and cel-miR-267 (control). Each sequence included mismatches at positions 10–12 to improve stability despite miRNA binding. PITA (http://genie.weizmann.ac.il/pubs/mir07/mir07_prediction.html) and RNAHybrid (http://bibiserv.techfak.uni-bielefeld.de/rnahybrid/) were used to optimize the sequences of the sponges. Algorithms predict the effectiveness of the designed binding site of the sponges by calculating free energy gained by binding to the miRNA and providing information on all other endogenous miRNAs that can potentially bind to the sponge sequence. This information was used to minimize off target miRNA binding and the sequences were inserted into pCMX-Myc-IRES2EGFP (gift from X. Morin, ENS, Paris, France) digested with XhoI and SalI. This construct allows miRNA sponge expression under pCAGGS promoter and was validated in a luciferase assay. Sponges were further subcloned into pNeuroD-IRES-GFP for in vivo experiments.

2.2 Antisense Oligonucleotides

AMOs (Integrated DNA Technologies, Leuven, Belgium), containing stabilizing 2′OMe and phosphorothioate modifications, were synthetized using perfect miRNA sequence complementarity. Furthermore, light-activated AMOs were generated through the incorporation of nucleobase-caged phosphoramidites as described by the Deiters laboratory [41]. The photocaged nucleotide 2′OMe (6-nitropiperonyloxymethyl)-caged uridine prevents miRNA-mediated gene silencing, until it is activated through 365–405 nm light-exposure, which induces photochemical removal of the caging groups leading to AMO-miRNA hybridization and inhibition of miRNA function.

2.3 In Vivo Light Activation of AMOs

pNeuroD-IRES-GFP and caged-AMO-22, -124, and -21 (used as a negative control) were in utero electroporated in E14 NMRI mice. Brains from E16 electroporated embryos were embedded in 3 % agarose and sectioned (300 mm) with a vibratome (VT1000S, Leica). Four hours after electroporation, slices were irradiated for 1 min with a mercury lamp (Nikon C-LHGFI Intensilight) and a DAPI filter cube for excitation (Nikon Al Eclipse Ti microscope, 40× objective, 50 % lamp power). Brain slices were cultured up to 24 h in semidry conditions (Millicell inserts, Merck Millipore), in a humidified incubator at 37 °C in a 5 % CO_2 atmosphere in wells containing neurobasal medium supplemented with 1 % B27, 1 % N2, and 1 % penicillin/streptomycin (Gibco, Life Technologies).

2.4 In Utero Electroporation

In utero electroporation was performed as described previously [38, 44], with minor modifications. Briefly, pregnant mice (14.5 days) were deeply anesthetized with isoflurane in oxygen carrier (Abbot Laboratories Ltd, Kent, UK). Uteri were exposed through a 1.5 cm-laparotomy. Embryos were carefully lifted using ring forceps through the incision and placed on humidified gauze pads. Plasmid DNA solution (2–4 μg/μl), prepared using EndotoxinFree plasmid purification kit (Qiagen Benelux B.V.), mixed with 0.05 % Fast Green (Sigma, St. Louis, MO) was injected through the uterine wall into the lateral ventricles using pulled borosilicate microcapillary and a Femtojet microinjector (VWR International). Electroporation was performed using an ECM-830 BTX square wave electroporator (VWR International) by discharging five electrical pulses at 35 V (50 ms duration) across the uterine wall at 1 s intervals using 5 mm platinum tweezers electrodes (CUY650P5, Sonidel, Ireland) with the positive side directed to the medial wall of the ventricle into which the DNA was injected. Uterine horns were then repositioned in the abdominal cavity and the abdomen wall and skin were sewed with surgical sutures. The pregnant mice were injected with buprenorphine (Temgesic®, Schering-Plough, Brussels, Belgium) and were kept on a warm plate (37 °C) for recovery. The whole procedure was complete within 45 min. Several days following surgery, pregnant mice were sacrificed by neck dislocation and embryos were processed for tissue analyses to study radial migration.

2.5 Focal Electroporation

Brains from E14 mouse embryos from Dicer$^{lox/lox}$ or NMRI mice were dissected in 0.1 M phosphate-buffered saline (PBS, pH 7.4) containing 0.25 mM glucose (PBS/glucose) and transferred into liquid 3 % low-melting agarose (37 °C, Bio-Rad) and then further incubated on ice for 1 h. Embedded brains were cut coronally into 300 μm slice with a vibratome (VT1000S, Leica). Slices were then subjected to ex vivo electroporation with sponge or AMOs for miR-22, miR-124 or cel-miR-67 (negative control miRNA) (5 μM). After cutting, brain slices were transferred onto 1 % low-melting agarose placed onto a petri dish square platinum plate electrode (Nepagene). We injected nucleotide solution with 0.05 % Fast Green (Sigma, St. Louis, MO) into the intermediate zone (IZ) of brain slices using a pulled borosilicate microcapillary and Femtojet microinjector. All steps were performed at 4 °C. Electroporations were performed by placing a cover square platinum plate electrode (Nepagene) onto the microinjected region. Square electric pulses (100 V, 5 ms) were injected five times at 1 s intervals using an electroporator (ECM 830, BTX). Micropipettes and electrodes were guided using a micromanipulator (WPI) under a stereomicroscope (Stemi DV4, Carl Zeiss). Slices were transferred onto Matrigel (BD bioscience, USA) after electroporation and cultured for 3 days in semidry conditions in a humidified incubator at 37 °C in a 5 % CO_2 atmosphere in wells containing Neurobasal medium supplemented with 1 % B27, 1 % N2, and 1 % penicillin/streptomycin.

2.6 Time-Lapse Imaging

For slice culture, brains from E16 embryos from Dicer$^{lox/lox}$ or NMRI mice electroporated 2 days earlier were embedded in 3 % agarose and sliced (300 μm) with a vibratome (VT1000S, Leica, Wetzlar, Germany). Brain slices were cultured up to 24 h in semidry conditions (Millicell inserts, Merck Millipore), in a humidified incubator at 37 °C in a 5 % CO_2 atmosphere in wells containing neurobasal medium supplemented with 1 % B27, 1 % N2, and 1 % penicillin/streptomycin (Gibco, Life Technologies, Ghent, Belgium). They were then placed in a humidified and thermoregulated chamber maintained at 37 °C on the stage of an inverted confocal microscope. Time-lapse confocal microscopy was performed with a Nikon A1 Eclipse Ti laser scanning confocal microscope. Images were taken with a 40× objective and 25 successive "z" optical plans spanning 50 μm every 30 min during 10 h. Sequences were analyzed using Image J [45].

2.7 Mouse Lines

Time-pregnant NMRI (Janvier labs, Saint Berthevin, France), Dicer$^{lox/lox}$ (M. Merkenschlager, Londres), and FoxG1Cre/+ (J.-M. Hébert, New York) mice back-crossed in MF1 genetic background were housed under standard conditions and treated according to guidelines of the Belgian Ministry of Agriculture in agreement with the European community Laboratory Animal Care and Use Regulations (86/609/CEE, Journal Officiel des Communautés Européennes L358, 18 December 1986).

3 Results and Discussion

Here, we discuss the different methods that are currently used for miRNA loss-of-function studies in the developing mouse cortex.

A loss of function approach is the ideal way to study the role of miRNAs of interest in vivo because it reveals functions that depend on physiological miRNA levels [46]. Gene regulation by miRNAs establishes a threshold level of target messenger mRNA below which protein translation is highly repressed. In this context, overexpression of exogenous miRNAs can result in repression of nonphysiological target mRNAs [46]. There are three general methods for miRNA loss-of-function studies: genetic knockouts, small molecule inhibitors [47–49], and miRNA sponges [42].

Dicer inactivation disrupts the maturation of miRNAs and targeting *Dicer* expression in cortical progenitors using the conditional knockout mouse line Foxg1$^{Cre/+}$; Dicer$^{lox/lox}$ embryos (Dicer cKO) [50, 51] resulted in a strong cortical phenotype [13, 39]. The positioning of projection neurons in the cortical wall was affected in Dicer cKO embryos. However, as miRNAs are also important to control the biology of cortical progenitors, we could not decipher whether miRNAs cell autonomously controlled neuronal migration. Indeed the cortical wall of Dicer cKO embryos showed loss of glial scaffold integrity, poor cell survival, progenitor proliferation, and specification defects that could secondarily impact on the migration of newborn neurons [39]. Therefore, we took another specific approach to promote the acute deletion of Dicer in newborn cortical neurons. For that purpose, we deleted *Dicer* in a cohort of postmitotic neurons by using in utero electroporation (IUE) of plasmids expressing the Cre recombinase together with GFP under the control the regulatory sequence of *NeuroD1* (NeuroD:Cre-GFP) or GFP only, as control (NeuroD:GFP). Three days after the ablation of *Dicer*, the targeted projection neurons were less efficiently reaching the CP (Fig. 1). Compared to control, only 37 % of neurons reached the CP (Fig. 2) and some neurons were stuck in deep layers (Fig. 1). Moreover, the contribution of miRNAs to radial migration was also exemplified by multipolar–bipolar cell transition defects, a critical step of radial migration [17]. Genetic ablation of *Dicer* led to accumulation of multipolar neurons that would normally repolarized to progress by locomotion on radial glial fibers (Fig. 3) [39].

Dicer deletion has been widely used, but it provides only information on global requirements of miRNAs. In order to find specific miRNA candidates, we used bioinformatics tools to predict miRNAs that would interact with genes required for radial migration of projection neurons. The selection of miRNA targets by predicting approaches and filtering approaches is important to obtain highly reliable data. We divided existing resources into three categories, including analysis of miRNA expression profiles (detecting miRNAs present in developmental cortex based on microarray), mRNA profiling (with or without Dicer expression), and miRNA target prediction approaches. Sequence complementarity, evolutionary conservation and free energy among miRNA-mRNA duplexes are the most common features to identify miRNA targets. The bioinformatic analysis coupled with wet lab experiments allowed us to

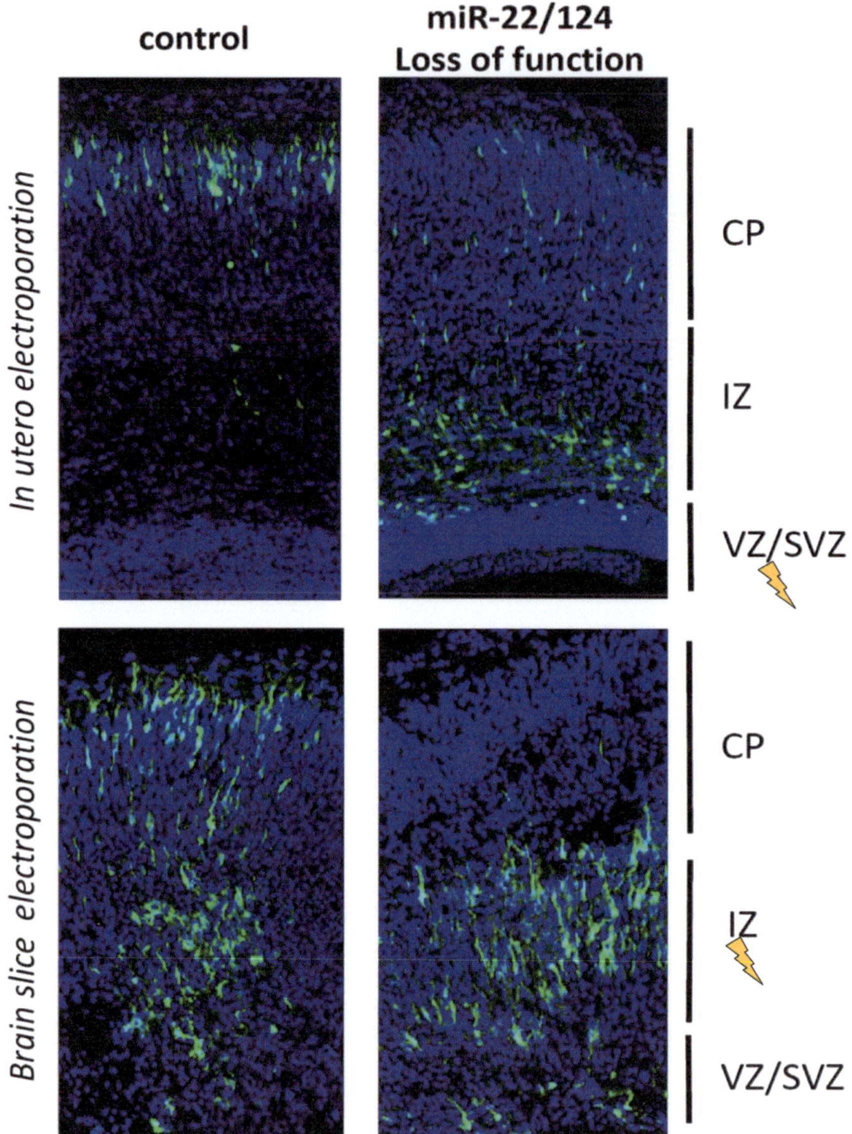

Fig. 1 Deletion of miR-22 and miR-124 leads to a defect of postmitotic neurons migration in mouse embryonic cortex. Loss-of-function experiments were performed using in utero and brain slice electroporation. *Lightning* indicated area of electroporation. *CP* cortical plate, *IZ* intermediate zone, *VZ/SVZ* ventricular/subventricular zones

identify miR-124 and miR-22 as regulators of projection neuron migration [39] (Fig. 1). By repressing CoREST expression which control Doulecortin expression, these miRNAs disrupt radial migration and thus control multipolar–bipolar cell transition in the IZ and bipolar stability during locomotion. In order to measure the contribution of these miRNAs to radial migration, we compared ex vivo and in vivo experimental approaches (Fig. 4).

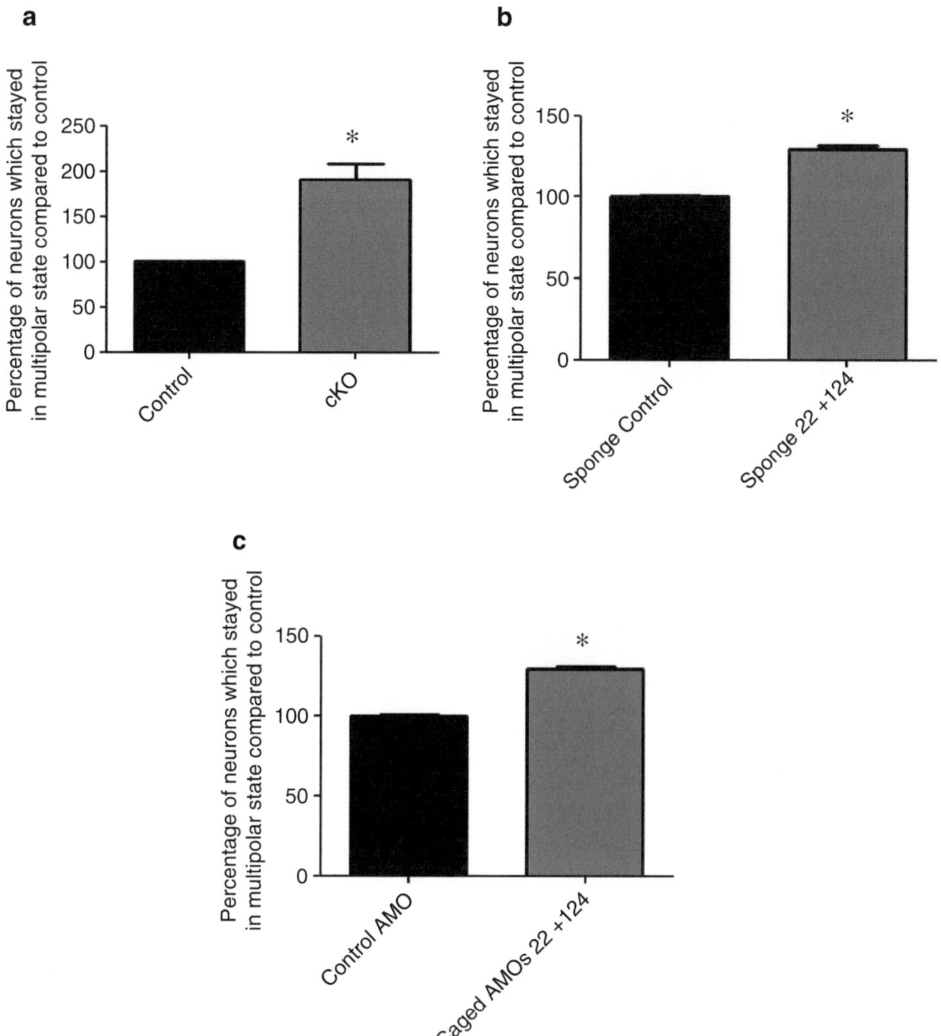

Fig. 2 AMOs and Sponge constructs lead to a defect in neuron migration similar to Dicer conditional KO. Acute depletion of Dicer (**a**) in Dicer$^{lox/lox}$ mice or acute depletion of miR-22 and miR-124 with specific sponge (**b**) or AMOs (**c**) in NMRI embryos impairs radial migration to upper layers. Cortical scattering of neurons was analyzed 3 days after electroporation of E14 embryos

Previous work showed how miRNAs control neuronal polarization during radial migration [39]. To compare the different approaches, we focused our attention on two criteria. We compared the percentage of: (1) migrating neurons which reach CP, and (2) multipolar–bipolar cell transition in the different experimental conditions.

A miRNA knockout mouse model aims induce genetic ablation of a specific miRNA. While deleting the gene coding a miRNA is the only way to guarantee complete loss of its activity, many miRNAs have seed family members encoded at multiple distant loci.

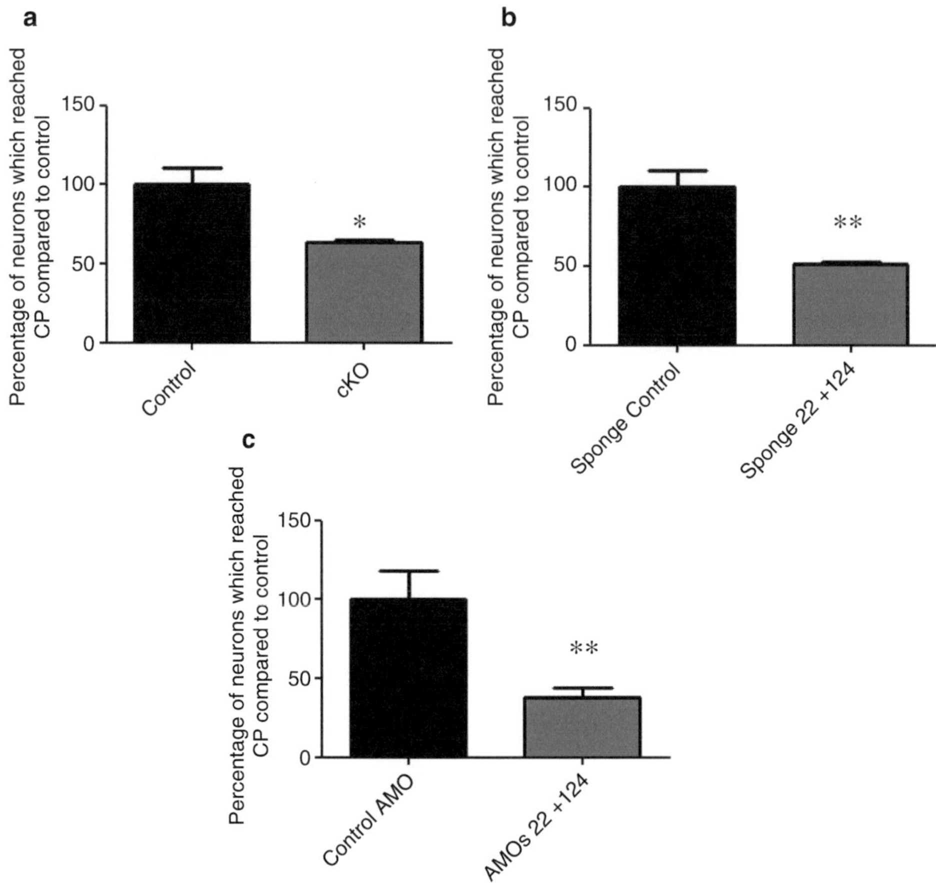

Fig. 3 Caged AMOs and Sponge constructs lead to multipolar stage persistence of postmitotic neurons with the same efficiency. Acute depletion of Dicer (**a**) in Dicer$^{lox/lox}$ mice or acute depletion of miR-22 and miR-124 with specific sponge (**b**) or AMOs (**c**) in NMRI embryos impairs bipolar–multipolar transition and lead to accumulation of multipolar cells throughout the cortical wall

According to the functional redundancy, these miRNAs would have to be knocked out individually and animals would need to be bred to generate the complete knockout strain of interest. An example of redundancy is demonstrated by *miR-133a-1* and *miR-133a-2*. Deletion of either *miR-133a-1* or *miR-133a-2* exhibits no obvious defects, whereas deletion of both results in late embryonic and neonatal death in approximately half of the mice with cardiac defects [52]. Moreover some miRNA precursors are transcribed in clusters; the proximity of the miRNAs within a cluster may render the specific deletion of one miRNA without affecting the processing of the others challenging. Moreover, one mRNA might be targeted by several miRNAs at different regions of the 3′UTR sequence. In this context, the analysis may be incomplete and it becomes difficult to shed light on the mechanisms involved. To overcome these problems other strategies were developed to study

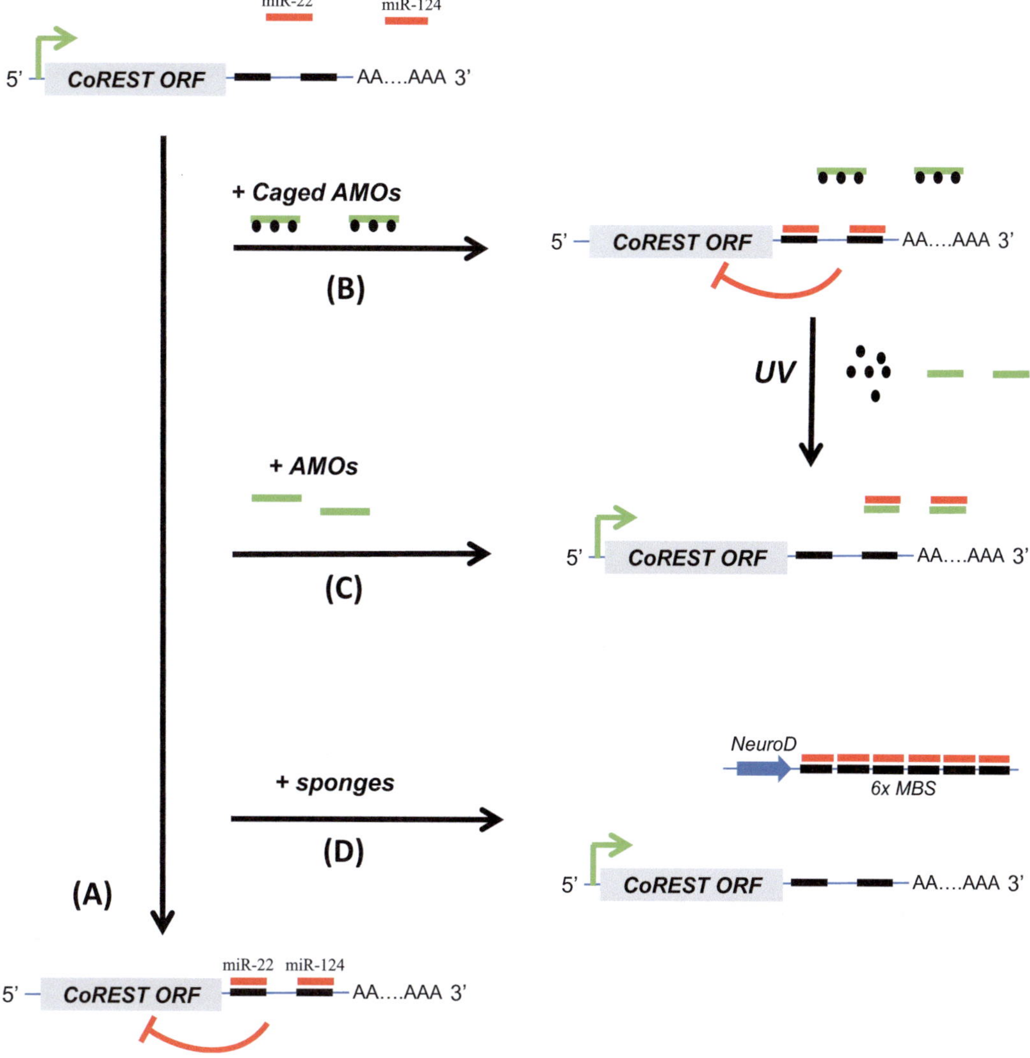

Fig. 4 miRNA silencing strategies used to characterize the role of miR-22/124 in cortical neurons. (*A*) In the absence of miRNA silencing, CoREST mRNAs are repressed by miR-22/124. (*B–D*) Introduction of the sponge transgene, AMOs or caged-AMOs, sequester the miRNA complexes, rescuing the expression of the endogenous targets and block their function in cortical neurons, leading to CoREST upregulation. (*B*) caged-AMOs need photochemical activation. UV light-activated caged-antagomiRs bind to endogenous miR-22 or miR-124

the role of a specific group of miRNAs in the developing mouse cortex.

Among these techniques, the culture of organotypic slices coupled to gene manipulation is attractive. This approach is suitable to screen gene function during neurodevelopment and may serve a very broad range of experimental paradigms. It allows the targeting of specific brain regions by precisely orienting the electrodes and it provides a high transfection efficiency of primary cortical progenitors. Brain slices survive in vitro for several days, and most importantly, recapitulate important milestones of cortical development, including radial migration [38]. Moreover, this technique allows functional manipulation and time-lapse recordings to monitor neuronal migration [39]. In this work, we acutely electroporated organotypic brain slices from E14 embryos with specific AMOs (Fig. 4). Since we assumed that our miRNAs of interest played a role in radial neuronal migration, electroporations were performed in the IZ where postmitotic neurons actively migrate. We also used a NeuroD:GFP construct to specifically follow postmitotic neurons in which we blocked the activity of endogenous miRNAs. After 3 days in vitro, we analyzed the radial migration of postmitotic neurons. Inhibition of miR-22 and miR-124 impaired the entry of electroporated neurons into the CP (Fig. 1). Compared to electroporation of a control AMO, only 38 % of neurons reached the CP and some neurons stuck in deep layers (Fig. 2).

In order to target miRNA in vivo, one can use in utero electroporation (IUE), a technique developed about 15 years ago [53–55] to investigate various aspects of cortical development [38]. IUE allows delivery of nucleotides sequences in vivo, including miRNA antagonist oligonucleotides, and subsequent functional analyses of genes through gain- and loss-of-function strategies. Despite being a powerful technique, IUE suffers from pitfalls, including the fact that: (1) electroporated molecules will primarily affect progenitors and not directly neurons, (2) the fine temporal and spatial regulation of endogenous miRNAs which imparts roles in brain morphogenesis [26, 27, 56–58] and neuronal cell fate [59–61] is bypassed by broad inhibition. Therefore, a defect in neuronal migration could indirectly result from targeting miRNA levels in progenitors and not from a specific function of these candidates in migrating neurons. To avoid this limitation, we used synthetic antagonist that could be light-activated in neurons migrating in organotypic slices. Optochemical control of miRNA inhibition has recently been achieved in cell culture with UV light-activated AMOs [41] (Fig. 4). This optopharmacological tool allowed us to induce spatial and temporal changes of endogenous miRNAs activity. For this purpose, cortical projection neurons from E14 embryos were co-electroporated in utero with NeuroD:GFP and caged AMOs that neutralize miR-22 and miR-124 after UV activation. Electroporated brains were sliced 2 days later and GFP areas in IZ were subjected to UV illumination. Real-time imaging was performed 24 h later and we analyzed the dynamic multipolar–bipolar conversion, a mechanism intimately linked to neuron migration. The number of multipolar neurons that became bipolar in control condition decreased, and involved a 30 % increase of the neurons that keep a multipolar shape after AMOs activation (Fig. 3). These results support the migration defect observed after acute slice electroporation.

Conditional cell targeting can be achieved by IUE of sponge constructs that act as dominant negative tools. They contain multiple target sites complementary to a miRNA of interest (Fig. 4). Sponges also offer advantages over chemically modified antisense oligonucleotide inhibitors. Indeed, antisense inhibitors require high sequence complementarity beyond the seed region of their target miRNA [18, 62]. Thus, the neutralization of a family of miRNAs may require the delivery of a mixture of perfectly complementary oligonucleotides. In striking contrast, when expressed at high levels sponges can inhibit the activity of a family of miRNAs sharing a common seed [32]). Moreover, regulatory elements can be added to the sponge promoter to make it drug-inducible or tissue-specific. In order to assess the role of endogenous miR-22 and miR-124 during cell shape remodeling of migrating projection neurons, we electroporated E14 embryos with NeuroD-driven miRNA sponges that neutralize endogenous miR-22 and miR-124 specifically in newborn projection neurons, and the analyses were conducted at E17. The neutralization of endogenous miR-22 and miR-124 by sponge constructs resulted in an impaired radial migration (Fig. 2). Indeed, compared to control, only 50 % of GFP neurons reach the CP. In order to check the polarity of migrating neurons, we also performed live imaging on cultured brain slices from E14 embryos electroporated with miRNA sponges and further harvested brains at E16. Expression of miR-22 and miR-24 sponge vectors impaired multipolar–bipolar neuronal conversion in the upper SVZ/lower IZ and increased the morphological instability of bipolar neurons locomoting in the upper IZ. The number of neurons that remain in a multipolar state increased by about 30 %. Similar data were obtained with electroporated and light uncaged-AMOs (30 %) (Fig. 3), however this frequency was lower compared to acute *Dicer* deletion. The apperent difference likely resulted from the overall disruption of miRNA maturation upon *Dicer* deletion as compared to expression of more specific sponges. However, taken together, the results lead to the same conclusion. Electroporation of those constructs impaired radial migration and promoted accumulation of multipolar neurons at the expense of bipolar ones.

Altogether, our experiments elucidated, with similar efficiency, the role of miR-22 and miR-124 in radial migration and multipolar–bipolar transition of postmitotic neurons. This indicates that, depending of the question, one can choose between different techniques to analyze the role of miRNAs in cortical development. When AMOs are used, the ex vivo electroporation approach is preferred because it allows targeting of specific regions of the brain by precisely orienting the electrodes. This approach is highly efficient to screen gene function during neurodevelopment and may serve a very broad range of experimental purposes. However, unless the slice technique is improved to sustain longer culture timings, these methods may not be ideal to assess events that occur at later developmental stage of development. On the other hand, embryos could reach adult stage following IUE, therefore allowing analyses of both the short- and long-term effects of genetic manipulation of miRNAs.

References

1. Bartel DP (2004) MicroRNAs: genomics, biogenesis, mechanism, and function. Cell 116 (2):281–297
2. Bartel DP (2009) MicroRNAs: target recognition and regulatory functions. Cell 136 (2):215–233
3. Lee Y, Ahn C, Han J, Choi H, Kim J, Yim J et al (2003) The nuclear RNase III Drosha initiates microRNA processing. Nature 425 (6956):415–419
4. Bernstein E, Caudy AA, Hammond SM, Hannon GJ (2001) Role for a bidentate ribonuclease in the initiation step of RNA interference. Nature 409(6818):363–366
5. Winter J, Jung S, Keller S, Gregory RI, Diederichs S (2009) Many roads to maturity: microRNA biogenesis pathways and their regulation. Nat Cell Biol 11(3):228–234
6. Fabian MR, Sonenberg N, Filipowicz W (2010) Regulation of mRNA translation and stability by microRNAs. Annu Rev Biochem 79:351–379
7. Friedman RC, Farh KK, Burge CB, Bartel DP (2009) Most mammalian mRNAs are conserved targets of microRNAs. Genome Res 19 (1):92–105
8. Friedman RC, Burge CB (2014) MicroRNA target finding by comparative genomics. Methods Mol Biol 1097:457–476
9. Bushati N, Cohen SM (2007) microRNA functions. Annu Rev Cell Dev Biol 23:175–205
10. Kawahara H, Imai T, Okano H (2012) MicroRNAs in neural stem cells and neurogenesis. Front Neurosci 6:30
11. Lang MF, Shi Y (2012) Dynamic roles of microRNAs in neurogenesis. Front Neurosci 6:71
12. Shi Y, Zhao X, Hsieh J, Wichterle H, Impey S, Banerjee S et al (2010) MicroRNA regulation of neural stem cells and neurogenesis. J Neurosci 30(45):14931–14936
13. Volvert ML, Rogister F, Moonen G, Malgrange B, Nguyen L (2012) MicroRNAs tune cerebral cortical neurogenesis. Cell Death Differ 19(10):1573–1581
14. Rakic P (2007) The radial edifice of cortical architecture: from neuronal silhouettes to genetic engineering. Brain Res Rev 55 (2):204–219
15. Hevner RF, Haydar TF (2012) The (not necessarily) convoluted role of basal radial glia in cortical neurogenesis. Cereb Cortex 22 (2):465–468
16. Germain N, Banda E, Grabel L (2010) Embryonic stem cell neurogenesis and neural specification. J Cell Biochem 111(3):535–542
17. Noctor SC, Martinez-Cerdeno V, Ivic L, Kriegstein AR (2004) Cortical neurons arise in symmetric and asymmetric division zones and migrate through specific phases. Nat Neurosci 7(2):136–144
18. Davis S, Lollo B, Freier S, Esau C (2006) Improved targeting of miRNA with antisense oligonucleotides. Nucleic Acids Res 34 (8):2294–2304
19. Kawase-Koga Y, Otaegi G, Sun T (2009) Different timings of Dicer deletion affect neurogenesis and gliogenesis in the developing mouse central nervous system. Dev Dyn 238 (11):2800–2812
20. Nowakowski TJ, Mysiak KS, Pratt T, Price DJ (2011) Functional dicer is necessary for appropriate specification of radial glia during early development of mouse telencephalon. PLoS One 6(8):e23013
21. Makeyev EV, Zhang J, Carrasco MA, Maniatis T (2007) The MicroRNA miR-124 promotes neuronal differentiation by triggering brain-specific alternative pre-mRNA splicing. Mol Cell 27(3):435–448
22. Bentwich I (2005) Prediction and validation of microRNAs and their targets. FEBS Lett 579 (26):5904–5910
23. Johnston RJ, Hobert O (2003) A microRNA controlling left/right neuronal asymmetry in Caenorhabditis elegans. Nature 426 (6968):845–849
24. Kloosterman WP, Plasterk RH (2006) The diverse functions of microRNAs in animal development and disease. Dev Cell 11 (4):441–450
25. Liang RQ, Li W, Li Y, Tan CY, Li JX, Jin YX et al (2005) An oligonucleotide microarray for microRNA expression analysis based on

labeling RNA with quantum dot and nanogold probe. Nucleic Acids Res 33(2):e17

26. Mansfield JH, Harfe BD, Nissen R, Obenauer J, Srineel J, Chaudhuri A et al (2004) MicroRNA-responsive 'sensor' transgenes uncover Hox-like and other developmentally regulated patterns of vertebrate microRNA expression. Nat Genet 36(10):1079–1083
27. Nelson PT, Baldwin DA, Kloosterman WP, Kauppinen S, Plasterk RH, Mourelatos Z (2006) RAKE and LNA-ISH reveal microRNA expression and localization in archival human brain. RNA 12(2):187–191
28. Schratt GM, Tuebing F, Nigh EA, Kane CG, Sabatini ME, Kiebler M et al (2006) A brain-specific microRNA regulates dendritic spine development. Nature 439(7074):283–289
29. Shibata M, Kurokawa D, Nakao H, Ohmura T, Aizawa S (2008) MicroRNA-9 modulates Cajal-Retzius cell differentiation by suppressing Foxg1 expression in mouse medial pallium. J Neurosci 28(41):10415–10421
30. Krek A, Grun D, Poy MN, Wolf R, Rosenberg L, Epstein EJ et al (2005) Combinatorial microRNA target predictions. Nat Genet 37 (5):495–500
31. Kruger J, Rehmsmeier M (2006) RNAhybrid: microRNA target prediction easy, fast and flexible. Nucleic Acids Res 34(Web Server issue): W451–W454
32. Lewis BP, Shih IH, Jones-Rhoades MW, Bartel DP, Burge CB (2003) Prediction of mammalian microRNA targets. Cell 115(7):787–798
33. Xie X, Lu J, Kulbokas EJ, Golub TR, Mootha V, Lindblad-Toh K et al (2005) Systematic discovery of regulatory motifs in human promoters and 3′ UTRs by comparison of several mammals. Nature 434(7031):338–345
34. Ma Y, Yao N, Liu G, Dong L, Liu Y, Zhang M et al (2015) Functional screen reveals essential roles of miR-27a/24 in differentiation of embryonic stem cells. EMBO J 34(3):361–378
35. Flemr M, Buhler M (2015) Single-step generation of conditional knockout mouse embryonic stem cells. Cell Rep 12(4):709–716
36. Park CY, Choi YS, McManus MT (2010) Analysis of microRNA knockouts in mice. Hum Mol Genet 19(R2):R169–R175
37. Creppe C, Malinouskaya L, Volvert ML, Gillard M, Close P, Malaise O et al (2009) Elongator controls the migration and differentiation of cortical neurons through acetylation of alpha-tubulin. Cell 136 (3):551–564
38. Nguyen L, Besson A, Heng JI, Schuurmans C, Teboul L, Parras C et al (2006) p27kip1 independently promotes neuronal differentiation and migration in the cerebral cortex. Genes Dev 20(11):1511–1524
39. Volvert ML, Prevot PP, Close P, Laguesse S, Pirotte S, Hemphill J et al (2014) MicroRNA targeting of CoREST controls polarization of migrating cortical neurons. Cell Rep 7 (4):1168–1183
40. Krutzfeldt J, Kuwajima S, Braich R, Rajeev KG, Pena J, Tuschl T et al (2007) Specificity, duplex degradation and subcellular localization of antagomirs. Nucleic Acids Res 35 (9):2885–2892
41. Connelly CM, Uprety R, Hemphill J, Deiters A (2012) Spatiotemporal control of microRNA function using light-activated antagomirs. Mol Biosyst 8(11):2987–2993
42. Ebert MS, Neilson JR, Sharp PA (2007) MicroRNA sponges: competitive inhibitors of small RNAs in mammalian cells. Nat Methods 4(9):721–726
43. Ebert MS, Sharp PA (2010) Emerging roles for natural microRNA sponges. Curr Biol 20(19): R858–R861
44. Creppe C, Malinouskaya L, Volvert ML, Close P, Laguesse S, Gillard M et al (2010) Elongator orchestrates cerebral cortical neurogenesis. Med Sci 26(2):135–137
45. Schneider CA, Rasband WS, Eliceiri KW (2012) NIH Image to ImageJ: 25 years of image analysis. Nat Methods 9(7):671–675
46. Mukherji S, Ebert MS, Zheng GX, Tsang JS, Sharp PA, van Oudenaarden A (2011) MicroRNAs can generate thresholds in target gene expression. Nat Genet 43(9):854–859
47. Krutzfeldt J, Rajewsky N, Braich R, Rajeev KG, Tuschl T, Manoharan M et al (2005) Silencing of microRNAs in vivo with 'antagomirs'. Nature 438(7068):685–689
48. Orom UA, Kauppinen S, Lund AH (2006) LNA-modified oligonucleotides mediate specific inhibition of microRNA function. Gene 372:137–141
49. Thomas M, Deiters A (2013) MicroRNA miR-122 as a therapeutic target for oligonucleotides and small molecules. Curr Med Chem 20 (29):3629–3640
50. Cobb BS, Nesterova TB, Thompson E, Hertweck A, O'Connor E, Godwin J et al (2005) T cell lineage choice and differentiation in the absence of the RNase III enzyme Dicer. J Exp Med 201(9):1367–1373
51. Hebert JM, McConnell SK (2000) Targeting of cre to the Foxg1 (BF-1) locus mediates loxP recombination in the telencephalon and other developing head structures. Dev Biol 222 (2):296–306

52. Liu N, Bezprozvannaya S, Williams AH, Qi X, Richardson JA, Bassel-Duby R et al (2008) microRNA-133a regulates cardiomyocyte proliferation and suppresses smooth muscle gene expression in the heart. Genes Dev 22 (23):3242–3254
53. Fukuchi-Shimogori T, Grove EA (2001) Neocortex patterning by the secreted signaling molecule FGF8. Science 294 (5544):1071–1074
54. Saito T, Nakatsuji N (2001) Efficient gene transfer into the embryonic mouse brain using in vivo electroporation. Dev Biol 240 (1):237–246
55. Tabata H, Nakajima K (2001) Efficient in utero gene transfer system to the developing mouse brain using electroporation: visualization of neuronal migration in the developing cortex. Neuroscience 103(4):865–872
56. Aboobaker AA, Tomancak P, Patel N, Rubin GM, Lai EC (2005) Drosophila microRNAs exhibit diverse spatial expression patterns during embryonic development. Proc Natl Acad Sci U S A 102(50):18017–18022
57. Giraldez AJ, Cinalli RM, Glasner ME, Enright AJ, Thomson JM, Baskerville S et al (2005) MicroRNAs regulate brain morphogenesis in zebrafish. Science 308(5723):833–838
58. Vo N, Klein ME, Varlamova O, Keller DM, Yamamoto T, Goodman RH et al (2005) A cAMP-response element binding protein-induced microRNA regulates neuronal morphogenesis. Proc Natl Acad Sci U S A 102 (45):16426–16431
59. Krichevsky AM, King KS, Donahue CP, Khrapko K, Kosik KS (2003) A microRNA array reveals extensive regulation of microRNAs during brain development. RNA 9 (10):1274–1281
60. Krichevsky AM, Sonntag KC, Isacson O, Kosik KS (2006) Specific microRNAs modulate embryonic stem cell-derived neurogenesis. Stem Cells 24(4):857–864
61. Smirnova L, Grafe A, Seiler A, Schumacher S, Nitsch R, Wulczyn FG (2005) Regulation of miRNA expression during neural cell specification. Eur J Neurosci 21(6):1469–1477
62. Esau CC (2008) Inhibition of microRNA with antisense oligonucleotides. Methods 44 (1):55–60

Neuromethods (2017) 128: 89–117
DOI 10.1007/7657_2016_3

Published online: 2 August 2016

Continuous Delivery of Oligonucleotides into the Brain

Ilya A. Vinnikov, Andrii Domanskyi, and Witold Konopka

Abstract

The growing field of RNA neurobiology dictates development and improvement of effective and reliable in vivo techniques to address the function of particular microRNA molecules within the brain. Here we describe a novel method involving continuous delivery of oligonucleotides into a brain region of interest by osmotic pump infusion. The approach implements application of double-stranded microRNA-mimics with only two LNA moieties at the 3′-end and additionally one at the 5′-end of the sense strand. This method holds promise for long-lasting and specific siRNA upregulation in vivo, especially in the Dicer-depleted systems, where other approaches are limited or not applicable. Being robust and effective, various techniques described in this chapter can be easily modified in order to achieve up- or downregulation of expression of specific RNA molecules, bi- or unilateral infusions or injections, and in vivo "screening" strategy allowing to start from a bigger group of RNA molecules and end up with identification of single RNA species critical for a phenotype.

Keywords: MicroRNA, Brain, Infusion, Neuroscience, Mice, Neuron

1 Introduction

The field of non-coding RNAs evolves rapidly. The understanding grows that reliable in vivo techniques are required to address the function of particular RNA molecules. MicroRNAs guide the silencing machinery to the target mRNA in order to suppress its translation and/or stability [1]. The ability of microRNAs to keep target mRNA translation silent until it is required makes them especially useful for regulation of translation in neurons due to the elaborated network of processes [2–4]. Previously, we and others have demonstrated that the loss of Dicer, a key nuclease for microRNA maturation, modifies neuronal activity in the brain, and hence, influences the processes of memory formation [5], appetite and weight [6], and others. Additionally, several individual microRNAs have been implicated in other aspects of brain physiology [7–9].

The important question arises how to link the global micro-RNA biogenesis pathway function with the effects of individual microRNAs and if there is a way to modulate microRNA expression in vivo.

1.1 Types of Molecules Used to Alter MicroRNA Expression

1.1.1 Differences in Function and Nucleotide Sequence

In order to increase or decrease the expression of specific microRNAs, researchers use either mimics or inhibitors which could be either injected to the tissue directly, be expressed in viral vectors [10] or be genetically inserted by means of CRISPR-Cas9 technique globally [11] or Cre-dependently [12]. In regard to nucleotide sequences, the inhibitors can be designed differently, for example as microRNA sponges, miRZips or tough decoys [10]. Such viral/oligonucleotide injections proved to be efficient in studies on the central nervous system of embryos, early postnatal [13] and adult rodents [6] and will be likely used in the nearest future also for higher mammals. The next level of specificity can be achieved by administration of target protectors which block the binding sites of microRNAs on 3′-UTRs of their predicted/validated targets. This method represents currently the only way to characterize and validate simultaneously both microRNA/target interactions and the impact of such interactions on physiological function/phenotype in vivo [14].

However, some scientific questions cannot be solved by expression of siRNA precursors.

1.1.2 Modifications of Oligonucleotides

For example, when working with systems with conditional ablation of Dicer, one may try to restore particular microRNAs that are missing in the system expecting to revert the phenotype. However, precursors of siRNAs delivered by viral vectors or genetically encoded by nuclease-mediated techniques require Dicer nuclease for their maturation and hence are inappropriate in cases when Dicer is ablated. That is why the direct injection or infusion of oligonucleotides represents the only alternative for such studies.

> Delivery ot oligonucleotides in vivo requires (1) their prolonged stability which (2) should not compromise their targeting function or (3) cause toxicity; (4) additionally, their structure should favor binding to specific targets thus decreasing the off-target effects.

Modifications of the Ribose Ring

Since the first in vivo delivery of antagomirs [15] and their subsequent intracranial implementation [16] by the group of Markus Stoffel, a decade of technical improvement and in-depth research resulted in a plethora of methods varying in the specificity, toxicity, stability, and cost of synthesis of oligonucleotides. Several studies proved that 2′-*O*-methyl- [15, 17], 2′-Fluoro- [18], 2′-*O*-methoxyethyl- [19, 20] and locked nucleic acid (LNA) [21–23] and other modifications of the ribose rings drastically increase the resistance to nucleases and hence the stability of these molecules in the tissue [24]. Sugar moieties of the nucleotides containing a methylene bridge between the 2′-oxygen and the

4′-carbon of the ribofuranose ring (locked nucleic acid, LNA) are very widely used in oligonucleotide design. The reason is on the one hand that the 3-endo conformation of the ribose ring in which this bicyclic structure is locked mimics the standard A form helical structure of the RNA molecule. Thus LNA is compatible with the intact siRNA machinery [25]. On the other hand, LNA locks a monomer in a rigid bicyclic N-type conformation which confers both exceptional stability against nucleases and extremely low cellular toxicity. Moreover, tight binding between LNA and RNA nucleotides (see below) may be used to provide preferential incorporation of a specific oligonucleotide chain into the RNA-induced silencing complex (RISC).

Modifications of the Phosphate Backbone

Another widely used modification called phosphorothioate bases implies substitution of a non-bridging oxygen in the phosphate backbone of the oligonucleotide by a sulfur atom [14, 16, 19]. Generally, this modification decreases affinity of the oligonucleotide to its target sequence. Nevertheless, precise empirical stoichiometric combination of the phosphorothioate backbone with 2′ modifications of the ribose ring which tend to increase affinity can result in very stable single-stranded oligonucleotides preserving their targeting function [26].

Sequence-Related Distribution of Modifications and their Impact on Functionality and Toxicity

Notably, excessive modifications result in decreased physiological efficiency and increased toxicity. Indeed, total 2′-*O*-methyl or LNA modification of ribose rings or phosphorothioate modifications of every second or all internucleoside linkages can render the oligonucleotides to complete loss of activity and increase cytotoxic effects [16, 27, 28] unless the precise positioning of the modified bases is performed [26].

However, even minimal LNA modifications of the 3′ overhangs of siRNAs can effectively stabilize siRNAs [6, 27] and provide a compromise between stability and functionality for both in vitro and in vivo applications [25, 27].

Specific Positioning of Modifications Can Reduce Off-Target Effects

Importantly, stabilization of the nucleotide pair involving the first 5′ nucleotide of the passenger strand by using e.g. an LNA moiety favors the incorporation of the guide strand to the RISC [25, 27]. Another way to strongly reduce the off-target effects is to use small internally segmented interfering RNAs [29]. This simple strategy implies splitting of the passenger strand so that only the guide strand is recruited for targeting.

Complexes of Oligonucleotides with 3D Structured Agents

Binding of chemically stabilized oligonucleotides to nanoparticles leads to more effective silencing in vivo [30] without evident immune response. Despite the studies showing that complexation with cationic liposomes may potently induce immune response [27], different liposomal formulations were shown to be effective in delivering the modified nucleotides into cells of interest [31, 32]. Moreover, a use of HiPerfect liposome reagent (Qiagen) showed a great potential in studies involving delivery of oligonucleotides to the brain tissue [6, 14].

Several groups came up with methodologies robust enough to be used in different in vivo conditions and addressing different biological questions [6, 22, 26].

1.2 Ways of Delivery and Pharmacokinetics

1.2.1 Phenotype Dynamics and Delivery Techniques

When planning experiments, researchers should take into account the dynamics of the expected effects. For example, acute phenotype could be modulated by injections of the oligonucleotides, chronic processes require long-lasting infusions, while some phenotypes require stable genetic modification (e.g. by means of in vivo gene-modifying techniques). Sometimes, two different techniques may even have comparable effectivity (e.g. unilateral infusion and bilateral injection) [6]. In this case the obvious recommendation is to use the less sophisticated and more robust technique (oligonucleotide injection procedure). Why is the infusion technique more demanding? Mainly, because it requires prolonged placement of cannulas into the tissue which inevitably causes inflammation, might impact survival of many cells and cause adverse effects on both expected phenotype and fitness of the animals. Generally, it may not be expected that none of the animals in the experimental group are free of any of such effects.

1.2.2 Sites of Delivery

Other important considerations include the target site of the intracranial injection/infusion. In some cases it is relevant to target a specific brain region directly [16], in others–to deliver to the cerebroventricular system [26], or even intranasally [33].

Concerning the direct deliveries into the tissue, one should be careful not to disrupt the brain regions of interest or the structures situated on the way of the delivery track. Sometimes, very tiny damage of brain nuclei may per se impact the studied phenotype or cause drastic worsening of the animal's condition and even death. In this regard, it is also very important to carefully choose and deliver a relevant volume and overall dose of the substance to assure robust microRNA pathway modulation within the target neuronal population without toxicity effects and with minimal impact on the neighboring cells.

1.2.3 Other Pharmacokinetics Issues

Cellular uptake of oligonucleotides can be enhanced by using liposomal formulations. As was mentioned previously, the use of chemically modified oligonucleotides decreases their nuclease degradation and their elimination from the tissue thus further increasing their bioavailability for the cells of interest. Other important pharmacokinetic effects require direct studies addressing such parameters as distribution both on the cellular and sub-cellular levels, stability, elimination, diffusion throughout the tissue from the site of delivery, and others.

1.3 Concluding Remarks

Recently, we have developed a novel method to address the function of specific neuronal microRNAs in vivo [6]. While single-stranded siRNAs function through several silencing mechanisms [18], we have used minimally LNA modified double-stranded oligonucleotides to better mimic the microRNA-mediated RNA-interference pathway.

This in vivo approach of microRNA restoration implements a delivery of liposomal formulation of double-stranded microRNA-mimics with only two LNA moieties at the 3′-end into specific brain structures by osmotic pump infusions. Prolonged infusion of the LNA microRNA-mimics but not scrambled oligonucleotides attenuated the Dicer-dependent morphological and behavior phenotypes in mice lacking microRNAs in specific neuronal populations [6]. This method holds promise for siRNA upregulation in vivo, especially in the Dicer-depleted systems, where genetic methods are not applicable, as the maturation of the pre-microRNA molecules still requires a cleavage by this enzyme. Conversely, delivery of microRNA inhibitors by means of osmotic pump-mediated infusion can be used to achieve a prolonged decrease of specific microRNA expression.

Universal strategy for intracranial treatment by oligonucleotides would fulfill the following criteria:

- Provide efficient delivery of microRNA mimics which must be
  - biologically functional,
  - stable in order to sustain long treatment periods,
  - easy to be incorporated into cells,
  - designed to decrease the off-target effects,
- Allow the use of both intracranial injections and infusions,
- Allow implementation of screening approaches by gradual decrease of the number of oligonucleotides in the injection mixture,
- Be applicable both for genetically intact and microRNA pathway-depleted systems (e.g. tissues with diminished function of Dicer/Drosha).

In Sections 3 and 4, we will describe several methods of infusion of LNA-modified double-stranded RNA molecules using osmotic minipump equipped either with metal cannula or glass micropipette. The following main parts will be discussed:

- **MicroRNA-mimics design**
- **Preparation of oligonucleotides for infusion**
- **Infusion setup preparation**
- **Operation**

2 Materials

2.1 Solutions and Other Liquid Materials

RNAse-free sterile saline (Medibena); artificial cerebrospinal fluid (ACSF); fast Green FCF (MP Biomedicals); HiPerfect transfection reagent (Qiagen); sterile saline for post-operative rehydration.

2.2 Oligonucleotides and Solid Materials

LNA-modified oligonucleotides with/without FAM-fluorescein moiety on the 3′-end (Exiqon); gel-consistency glue Loctite 454 (Loctite); light-sensitive glue PL 5151 (Best-Klebestoffe); stereotaxic cannula and minipump holder (Stoelting); osmotic pumps, for example Model "1004" providing a flow rate of 0.11 μl/h over 4 weeks (Alzet); infusion kit3 or polypropylene tubing (Alzet); surgical instruments, wedged filter paper sponge bars.

In addition, bilateral infusions using metal cannula setup require double connector (Plastics1), whereas for glass micropipette setup, T-connector with inner diameter 1 mm (Neolab); thick-wall borosilicate glass B100-58-15 (Sutter Instruments) or beveled, fire polished and 90 °-bent glass micropipettes (Biomedical Instruments), and diamond pencil are needed.

2.3 Devices

Stereotaxic frame; thermal cycler (preferably with ramp function); warming pad for surgical operations on rodents; mini drill with cut-off disc; high-intensity directed light source device for light-sensitive glue, and, if using glass micropipettes, Sutter Instruments Flaming/Brown micropipette puller model P-1000, P-97, or P-87.

3 Methods

3.1 Design of MicroRNA-Mimics with Minimal LNA-Modifications

1. The sequence of the antisense (guide) strand (Fig. 1a) is designed to be identical to the corresponding microRNA (consisting from intact RNA-nucleotides except for the 3′-overhangs, see below).
2. The 3′-end overhang of the antisense oligonucleotide is replaced by two thymine LNA-moieties to stabilize the mimic against 3′ exonucleases (25).
3. The sequence of the sense (passenger) strand is made fully complementary to the antisense strand (RNA-nucleotides except for the first 5′-nucleotide and the 3′-overhangs) thus increasing stability of the annealed double-stranded oligonucleotide against endonucleases.
4. The first 5′-sugar of the sense strand is modified by locking (LNA) thus allowing preferential recruitment of the antisense strand to the RISC [25, 27].

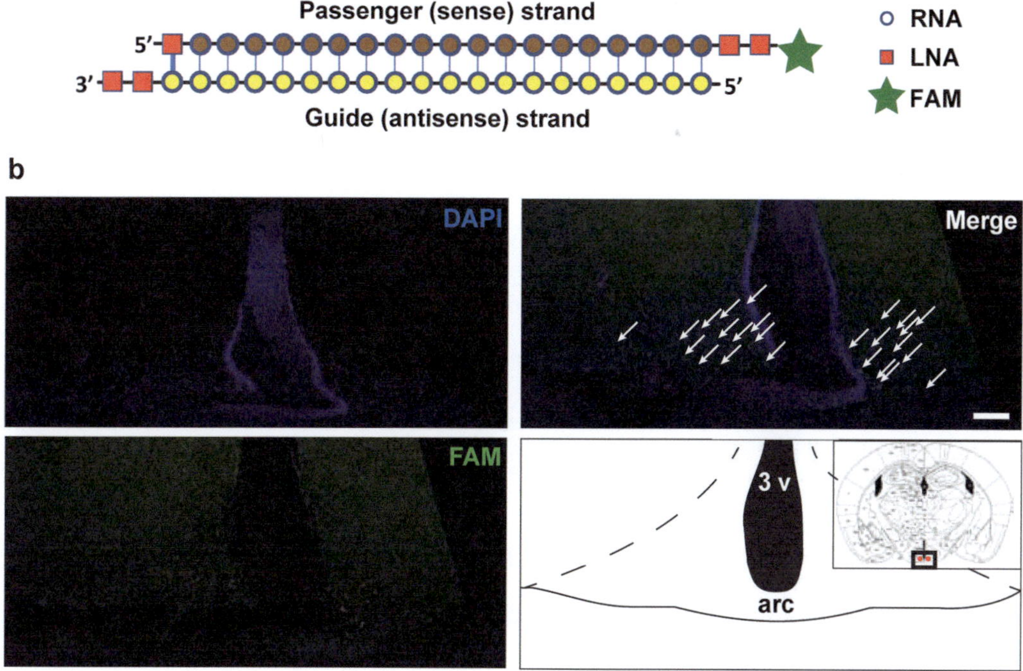

Fig. 1 (**a**) Design of microRNA-mimics. Note that oligonucleotides are only minimally modified by LNA moieties. LNA, locked nucleic acid; FAM, 3′(6-carboxy-fluorescein), 1-(4,4′-Dimethoxytrityloxy)-3-[*O*-(*N*-carboxy-(di-*O*-pivaloyl-fluorescein)-3-aminopropyl)]-propyl-2-*O*-succinoyl moiety. (**b**) Fluorescent microphotograph of the ventral part of the hypothalamus showing distribution and stability of minimally LNA-modified oligonucleotides in the mouse arcuate nucleus of the hypothalamus (*arc*). *Red dots* in the scheme [36] in the *inset* outline the injection sites. The signal (*arrows*) represents direct FAM fluorescence 2 weeks after bilateral injection of ACSF/liposomal formulation of FAM-labeled oligonucleotides (final concentration 1.44 μg/μl injected in a volume of 0.5 μl per side) to the following stereotaxic coordinates relative to bregma in mm: A/P −1.46, M/L ± 0.25, D/V −5.75. 3 v, third ventricle of the brain. Scale bar in μm: 100

5. The 3′-end overhang of the sense oligonucleolide is replaced by two thymine LNA-moieties to stabilize the mimic against 3′ exonucleases.
6. The last thymine LNA-nucleotide of the sense strand may be coupled with FAM-fluorescein moiety to achieve direct fluorescence or immunofluorescence-based evaluation of oligonucleotide distribution and pharmacokinetics (Fig. 1b).

See **Note 4.1**, Table 1 and Fig. 1 for details.

3.2 Preparation of Oligunoculeotides for Delivery

3.2.1. Resuspend the oligonucleotides in sterile RNAse-free 0.9 % NaCl.

3.2.2. Anneal the oligonucleotides:

3.2.2.1. *Mix equimolar amounts of both complementary oligonucleotides in a 100 μl sterile PCR microfuge tube.*

Table 1
Sequences of scrambled microRNA-mimic oligonucleotides

Oligonucleotide	Sequence	ID
Scrambled antisense	rUrGrGrGrCrGrUrArUrArGrArCrGrUrGrUrUrArCrArC **+T** **+T**	137099
Scrambled sense	**+G**rUrGrUrArArCrArCrGrUrCrUrArUrArCrGrCrCrCrA **+T** **+T**	137096
Scrambled sense (FAM)	**+G**rUrGrUrArArCrArCrGrUrCrUrArUrArCrGrCrCrCrA **+T** **+T** -FAM	N/A

All nucleotides are RNA, except the ones highlighted in bold and red and marked with "+" on the front, which represent LNA moieties. FAM stands for 1-(4,4′-Dimethoxytrityloxy)-3-[*O*-(*N*-carboxy-(di-*O*-pivaloyl-fluorescein)-3-aminopropyl)]-propyl-2-*O*-succinoyl (or 3′(6-carboxy-fluorescein)) moiety

3.2.2.2. *Place the tubes to a thermal cycler and set up a program to perform the following profile:*

1. *Heat the block to 95 °C and pause the program.*
2. *Place the tubes to the cycler block, resume the program and incubate at 95 °C for 2 min.*
3. *Ramp cool to 25 °C over a period of 45 min.*
4. *Proceed to a storage temperature of 4 °C.*
5. *Briefly spin the tubes in a microfuge to draw all moisture from the lid.*

3.2.3. Store on ice until ready to use.

3.2.4. Prior to use: supplement with the appropriate volume of ACSF to obtain the desired final concentration (consider also the step below).

3.2.5. Add 1.35 μl HiPerfect reagent for each 10 μl of the (ACSF + oligonucleotides) solution.

3.2.6. Mix intensively (vortex) for 1 min and leave at room temperature for 5 min before filling the pump.

See **Note 4.2** for more details.

3.3 Fabrication of Micropipettes and Assembly of the Infusion Setup

As mentioned above, two different strategies could be used to perform intracranial infusion of oligonucleotides: using metal cannulas (connectors) or glass micropipettes. Both of the strategies have their pros and cons. The major benefit of the metal cannulas is the possibility to achieve bilateral infusions, while the major benefit of the glass micropipettes is their safety due to a small size of the taper and drastically decreased electrical interactions with neuronal tissue.

3.3.1. *Setup involving a connector with a metal cannula*

3.3.1.1. *Assemble the connector with the polypropylene tubing and continue further from* ***step 3.3.3.1***

3.3.1.2. *General considerations*

The ease of the assembly of the infusion setup using unilateral or bilateral connectors

(Plastics1), as well as robustness of surgical intervention makes this way of intracranial delivery of substances very attractive (Fig. 2a). Importantly, the possibility to infuse bilaterally further strengthens the advantages of such method. As can be seen in the Fig. 2b, osmotic minipumps provide continuous delivery of substances intracranially giving a possibility to target large brain structures.

3.3.1.3. *Functional studies*

In the experiment illustrated by Fig. 2c we used the setup involving bilateral cannulas allowing infusion to adult mice into the region of the arcuate hypothalamic nucleus (stereotaxic coordinates relative to bregma in mm: A/P −1.46, M/L ±0.4, D/V −5.5). The mice get obese several weeks after tamoxifen-inducible Cre-dependent tissue specific knock-out of *Dicer1* gene in the forebrain neurons (including the arcuate hypothalamic nucleus). The hypothesis being tested was that one of several predicted microRNAs was responsible for the obesity phenotype [6]. Therefore, the mice were infused with a mixture of ten microRNAs to attenuate the obesity phenotype (Fig. 2c).

3.3.1.4. *Survival issues*

The main drawback of the experiments with metal cannulas is their strong impact on survival of the operated animals. It turned out that even installation of the cannulas to different brain regions without subsequent infusion would lead to decreased fitness and survival. Moreover, some regions turned out to be more vulnerable to such infusions than others (Fig. 2b, Table 2), even when the cannulas were placed unilaterally. After performing the above mentioned tests and experiments we concluded that even if one loses the effectivity of the method by injecting the oligonucleotides only unilaterally which weakens the extent of the phenotype rescue, it is preferable to use less harmful and more robust technique such as the one implementing glass micropipettes (see below).

3.3.2. *Setup involving glass micropipettes*

One may prefer using beveled fire-polished 90 °-bent micropipettes with outer/inner diameter of 100/58 μm (Biomedical Instruments) (Figs. 3 and 4) or clean break

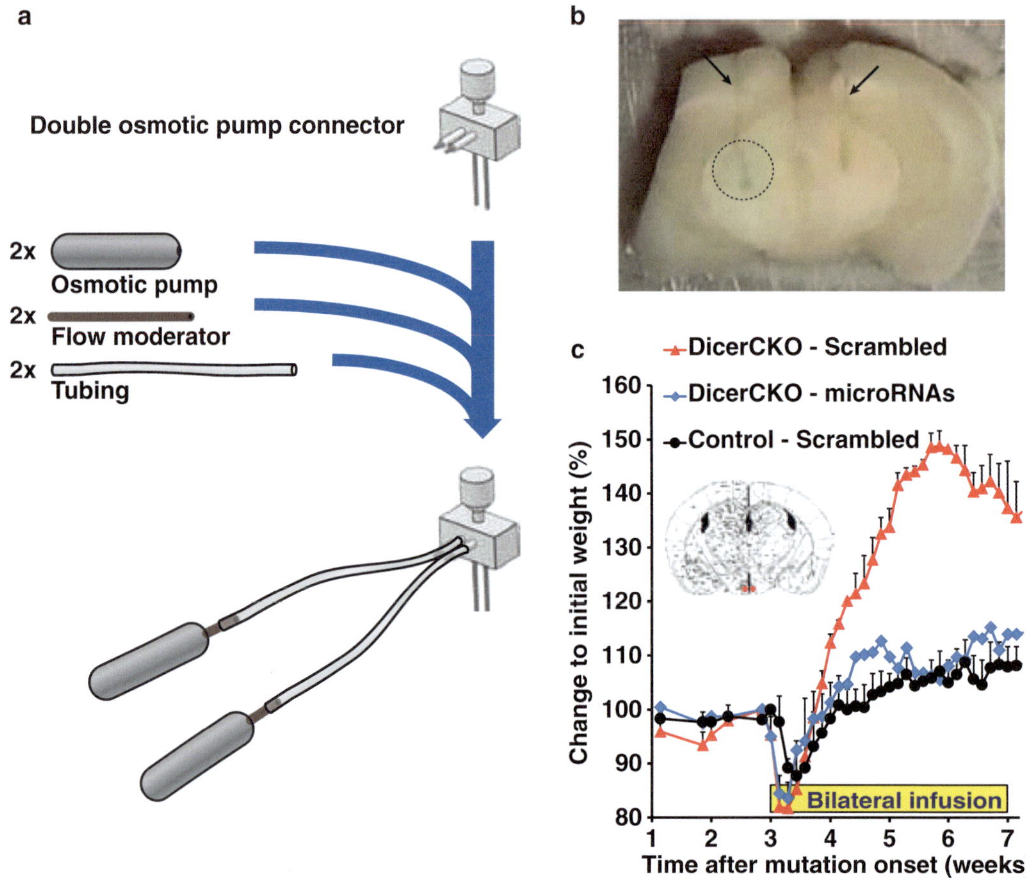

Fig. 2 (**a**) Schematic illustration of infusion system assembly implying double osmotic pump connector (not to scale). Flow moderator is depicted not fully inserted into the pump for better visual representation. Note that two sets of standard osmotic pump components are required for each double-cannula setup. (**b**) Microphotograph of a vibratome-cut brain section 3.5 weeks after unilateral infusion of the Fast Green FCF into the midbrain by means of double osmotic pump connector. *Dashed line* outlines the extent of diffusion in one of the hemispheres. Note very thick track from metal cannulas (*arrows*) severely destructing various brain structures during bilateral infusion (the ipsilateral side was infused with ACSF only). Stereotaxic coordinates relative to bregma in mm: A/P −3.08, M/L ±1.4, D/V −4. (**c**) Dynamics of body weight of controls and Dicer^{CaMKCreERT2} mice infused bilaterally to the arcuate hypothalamic nucleus with scrambled, as well as Dicer^{CaMKCreERT2} mice infused with a mixture of ten microRNA mimics predicted to target the mTOR pathway which has been found to be crucial for the obesity phenotype [6]. $n = 5$, of which 2, 2 and 1 survived, respectively, due to the use of metal cannulas. Final concentration of the oligonucleotides was 1.35 μg/μl. *Red dots* in the schematics [36] outline the sites of infusion (stereotaxic coordinates relative to bregma in mm: A/P −1.46, M/L ±0.4, D/V −5.5) by double osmotic pump connector

straight thick-wall borosilicate micropipettes B100-58-15 (Sutter) (Fig. 5) with a taper diameter of 50±5 μm to assemble the infusion setup. Both types are much less destructive for any brain region tested (Fig. 6) and substantially increase survival of the operated animals (Table 2). The following protocols are based on the studies of

Table 2
Survival rates after infusions into different brain structures

Area of implantation	Cannula type	Number of cannulas implanted (uni- or bilateral)	Infusion period (weeks)	Coordinates relative to bregma (in mm) Anterior/ posterior	Lateral	Dorsal/ ventral	Number of mice for the experiment	Of which survived and healthy at least 1 week following operation (not losing weight, not exhibiting rotations etc.)	Survived and healthy (%)
Arcuate hypothalamic nucleus	Metal	1	No infusion	−1.46	−0.4	−5.5	4	3	75
		2	4	−1.46	±0.4	−5.5	21	6	28.6
	Glass	1	4	−1.46	−0.4	−5.5	50	42	84
Xiphoid thalamic nucleus	Metal	1	4	−1.46	0	−4.5	18	12	66.7
Dorsomedial hypothalamic nucleus	Metal	1	4	−1.46	−0.75	−5	5	3	60
Ventro-medial hypothalamic nucleus	Metal	1	2	−1.46	−0.5	−5.3	6	1	16.7
Midbrain	Metal	2	4	−3.08	±1.4	−4	33	30	90.9
		1	2	−3	0	−4.5	8	3	37.5
Lateral ventricle	Metal	1	2 weeks	−0.34	0.9	−2.5	36	36	100

Note that the coordinates for infusions to the arcuate and dorsomedial hypothalamic nuclei were located more dorsally and laterally to prevent unintended disruption of the nuclei

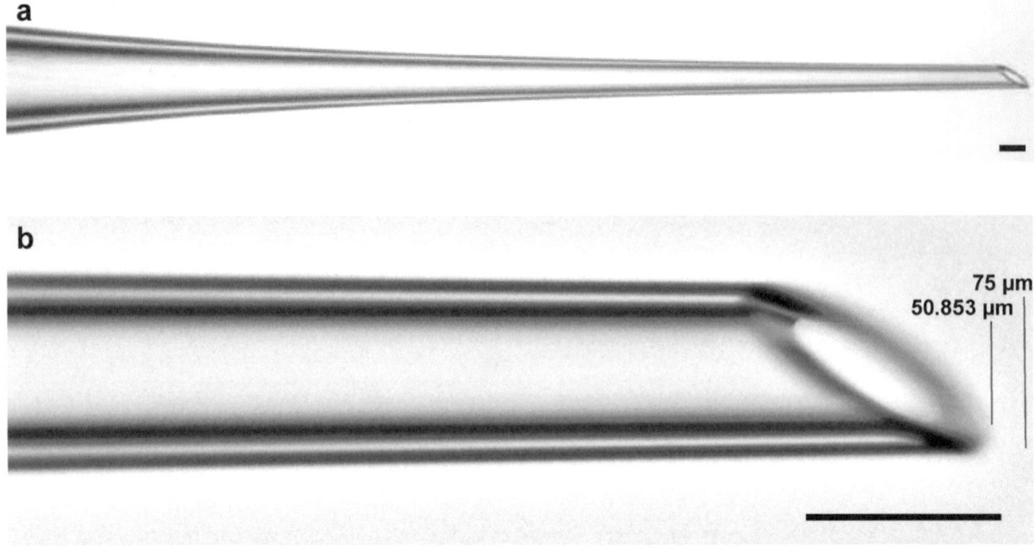

Fig. 3 (**a**) Gradual decrease of diameter of glass micropipette (Biomedical Instruments) towards the taper diameter of 50±5 μm. (**b**) Beveled and fire-polished taper of the same micropipette to reduce damage to the tissue. Scale bars in μm: 100

Cunningham et al. [34] but involve further development and modifications of this technique.

3.3.2.1. *Fire-polished beveled 90 °-bent micropipettes infusion setup*

1. Cut the bent micropipette (Biomedical Instruments) carefully approximately 1 cm away from the shoulder by a diamond pencil (Fig. 4b).
2. Cut both "arms" of the T-connector leaving only the "trunk."
3. Insert the thick end of the micropipette into the T-connector, place the "trunk" parallel to the axis of the pulled part of the micropipette and opposite to the taper and stabilize it with light-sensitive glue.
4. (optionally) Insert a pulled end of the micropipette through a flow moderator cap or a spacer from the Infusion kit3 and stabilize it perpendicularly with light-sensitive glue (Figs. 4b and 5c).
5. Insert the back end of the micropipette into polypropylene tubing and apply glue on the place of contact to ensure stable connection of both parts.

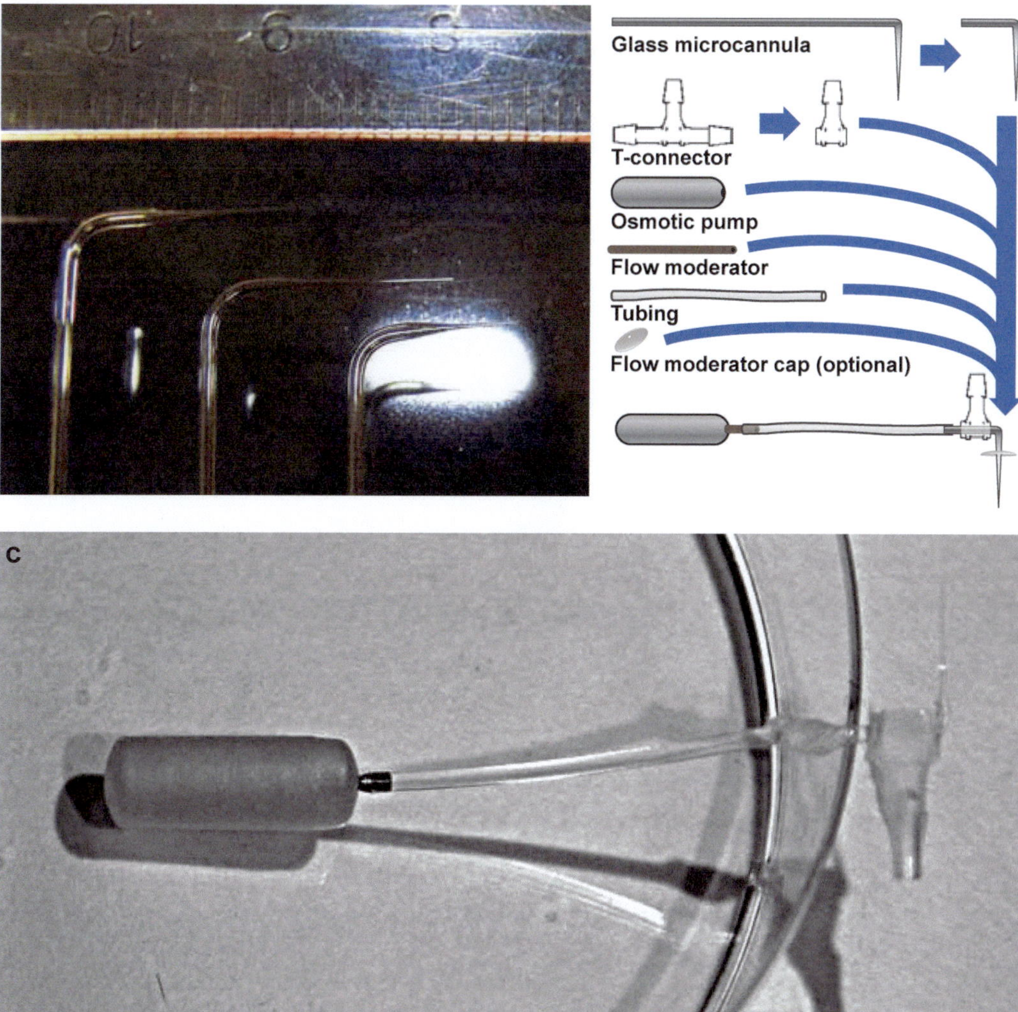

Fig. 4 (**a**) Photograph of 90 °-bent glass micropipettes (Biomedical Instruments). (**b**) Schematic illustration of infusion system assembly implying 90 °-bent micropipette (not to scale). Flow moderator is depicted not fully inserted into the pump for better visual representation. (**c**) The whole infusion setup implying 90 °-bent glass micropipette with fire-polished beveled taper. Note that flow moderator cap/insufision kit3 spacer is optional for such setup as a relatively big area of the bottom part of the T-connector provides stable enough contact with the skull

3.3.2.2. *Infusion setup using straight micropipettes, 50 μm tip with clean break*

This subtype of the glass micropipette requires relatively more time and effort to produce, so it is highly recommended to follow the protocol below. The first step involves pulling a pipette in a way that the capillary glass is heated and pulled

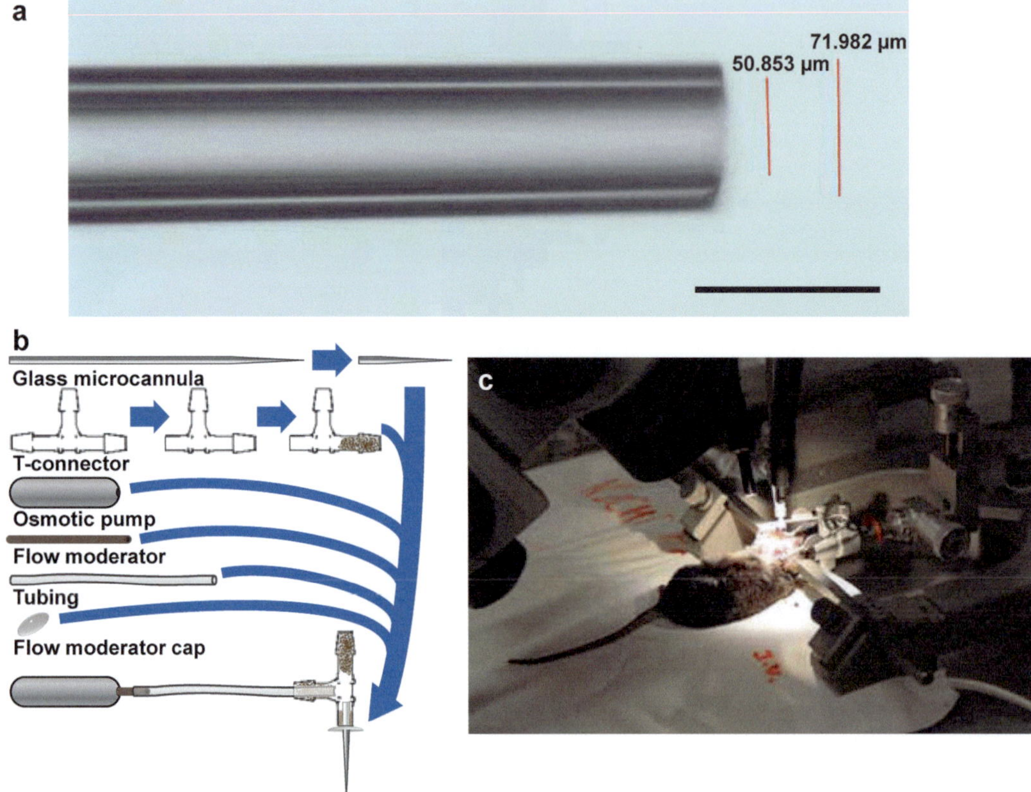

Fig. 5 (**a**) Glass-on-glass technique-broken back taper (inner diameter of the tip 50 μm) of straight thick-wall borosilicate glass micropipette B100-58-15. Scale bar in μm: 100. (**b**) Schematic illustration of infusion system assembly implying straight glass micropipette (not to scale). Flow moderator is depicted not fully inserted into the pump for better visual representation. Light-sensitive glue deposits are depicted in *brown*, however not at the place of a contact between tubing and flow moderator (for clarity). (**c**) Infusion setup during implantation and surgery procedure. The microscope was fixed on multidimensional freedom of movement arm allowing a view from various angles. This is critical for finding correct bregma-related coordinates and implantation as spacer/flow moderator cap restrict viewing the end of micropipette from the close to the upright positions of the microscope

to make a long, gradual taper. This is critical for infusions over long periods of time because the lumen of the micropipette inside the brain can become clogged. The second step is making a clean 90 ° break so that the tip does not have jagged edges, and thus does not damage the tissue. And the third step involves assembly of the micropipette with the T-connector and tubing into a tightly sealed system using light-sensitive glue.

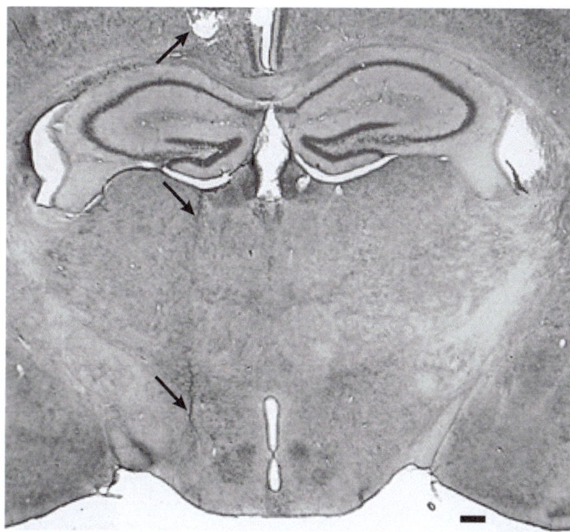

Fig. 6 Nissl staining of the brain after unilateral infusion into the arcuate hypothalamic nucleus (stereotaxic coordinates relative to bregma in mm A/P −1.46, M/L ±0.4, D/V −5.5). Micropipette track is indicated by *arrows*. Scale bar in μm: 100

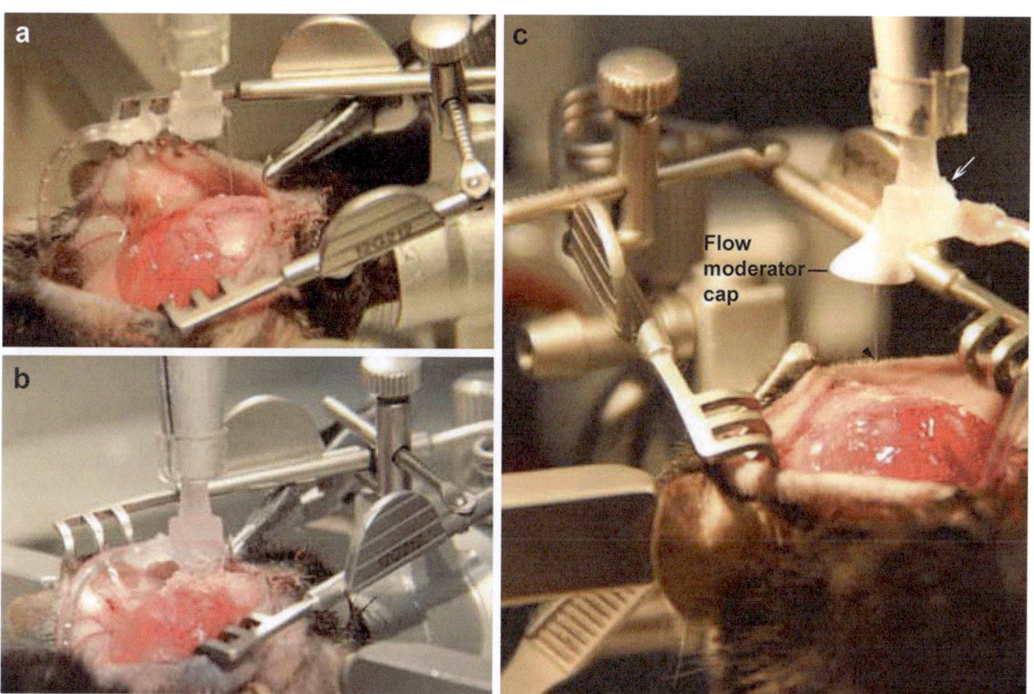

Fig. 7 Glass micropipette infusion setup implantation procedure. (**a**) Identification of correct coordinates before implantation. Osmotic pump is already introduced to the subcutaneous pocket before fixation of the T-connector by the stereotaxic cannula and minipump holder. Note silicon tubing on the branches of the holder to allow gentle but robust fixation of the T-connector. (**b**) Fixed infusion setup on the skull. Note the operation area on the skull is broad enough to exclude contact of the gel-consistency glue with skin which is critical especially for later closing the wound. (**c**) The infusion setup with flow moderator cap before implantation. The taper of the glass micropipette is indicated by an *arrowhead* and T-connector by an *arrow*

3.3.2.2.1. *Manual one-sided approach to pulling the micropipette*

CAUTION!!! The manual one-sided approach requires working with the humidity control chamber removed. While holding the left puller bar in place, the right puller bar is manually pushed outward to create a very long tapered pipette. Please contact Sutter Instrument prior to performing this procedure for detailed instructions. If not executed properly, the described technique may cause finger burns or damage to the filament. Please proceed at your own risk!

1. Install FB330B or FB255B filament to the puller.
2. Use Program #65 type E (http://www.sutter.com/PDFs/pipette_cookbook.pdf) using B100-58-15 glass and the following parameters:
   - Heat = Ramp temperature + 40
   - Pull = 0
   - Vel = 150
   - Time = 100
   - Pressure = 200
3. Install the capillary glass into the puller and keep the acrylic cover open and have the humidity control chamber removed. Make sure you wear protective glasses.
4. Hold the left pulled bar inward with your index finger using the finger hold and press <PULL>.
5. As the filament heals up and softens the glass, wait approximately 7 s (you might see/feel the right pulled bar to drift slightly to the right).

6. Once the glass has softened, slowly but continuously push the right pulled bar to the right while keeping the left pulled bar in place.
7. The long tapered pipette on the right side is the one on which a clean 90 ° break will be made in the next step.

3.3.2.2.2. *Producing a clean 90° break*

1. Please find the detailed information about the glass-on-glass technique developed by Adair L. Oesterle in the Pipette Cookbook (Sutter Instruments, http://www.sutter.com/PDFs/pipette_cookbook.pdf, pp. 59–60). Briefly, a pair of pipettes are pulled and one is used to score and break the other to produce a smooth clean 90 ° break. Alternatively, a CTS Ceramic Tile (Sutter Instrument) can be used to facilitate the scoring and breaking of the glass. The glass taper is scored approximately one-third back from the tip (or two-third up from the shoulder of the taper) to create a 50 μm tip inner diameter. Typically, this can result in the taper length of approximately 1 cm.

3.3.2.2.3. *Assembling the setup*

1. Cut the bent micropipette carefully ~0.5 cm longer away from the taper than the intended depth of the infusion by a diamond pencil (Fig. 5b).
2. Insert a pulled end of the micropipette through a flow moderator cap or a spacer from the Infusion kit3 and stabilize it perpendicularly with light-sensitive glue.

3. Cut one "arm" of the T-connector.
4. Insert the other, back end of the micropipette to the cut "arm" of the T-connector and stabilize it with light-sensitive glue.
5. Apply a small droplet of light-sensitive glue to the other arm and be ready to apply light as soon as it fills it completely.
6. Apply light-sensitive glue on the outer surface of the polypropylene tubing, insert it deep into the "trunk" of the T-connector and stabilize the assembly (Fig. 5b).

3.3.3. *Osmotic pump assembly protocol*

3.3.3.1. Attach flow moderator (Figs. 2a, 4b, c, and 5b) to the tubing preassembled with the cannula or micropipette (see the previous **steps 3.3.1.1**, **3.3.2.1.5**, or **3.3.2.2.3.6**), glue the connection to ensure robustness.

3.3.3.2. Fill the infusion setup from the previous step with a freshly made oligonucleotides mix supplemented with HiPerfect reagent (**step 3.2.6**).

3.3.3.3. Fill the pump with the mix using a special needle provided by the manufacturer, keep the pump vertically.

3.3.3.4. Slowly insert the flow moderator (from **step 3.3.3.2**) into the filled pump and use directly for surgery.

3.3.3.5. (optionally) If priming of the pumps is desired, leave the pump in sterile saline at 37 °C in a sealed Petri dish or falcon tube overnight and then use for surgery.

See **Note 4.3**, Table 2 and Figs. 2, 3, 4, 5, 6, and 7 for details.

3.4 Surgery

3.4.1. After anesthetizing the animal, open the skin over the skull (Figs. 5c and 7).

3.4.2. Make a pocket for the pumps by gently inserting the blunt ends of the forceps from the opening towards the back.

3.4.3. Fix the edges of the skin and dry the skull.

3.4.4. Make the hole/holes in the skull according to the coordinates relative to bregma.

3.4.5. Take fully equipped infusion setup and gently insert the pump into the pocket.

3.4.6. Use stereotaxic minipump holder to grab the arm of the infusion setup (*see* also **Note 4.3**).

3.4.7. Find the bregma with the taper of the micropipette and move to the final coordinates.

3.4.8. Carefully dry wide area of the skull under the taper again.

3.4.9. Insert the micropipette into the brain approximately halfway into the skull (so that the distance between the skull and the bottom surface the flow moderator cap/spacer/T-connector/base of the connector is around 3–5 mm) and use a tiny spatula to distribute 0.1–0.25 cm gel-consistency glue on the bottom surface.

3.4.10. Insert the pipette to the desired depth.

3.4.11. Wait 10–15 min.

3.4.12. Gently insert forceps or any other thin instrument between the sides of the holder and constantly apply moderate pressure on both sides by turning the instrument to open the branches of the holder while slowly unscrewing and then move up the holder.

3.4.13. Release the mouse from the stereotaxic frame.

3.4.14. Remove the arm of the cannula with scissors or, preferably, using mini drill with a cut-off disc.

3.4.15. Close the wound and leave the mouse to wake up on the warm pad.

See **Note 4.4**, and Figs. 5c and 7 for details.

4 Notes

4.1 MicroRNA-Mimic Design (Refers to Section 3.1)

Scrambled control oligonucleotides (Table 1) are minimally modified by LNAs as described in the Section 3.1 and Fig. 1a and have no homology to any known mammalian gene.

The Fig. 1b demonstrates that oligonucleotides injected to the arcuate hypothalamic nucleus (stereotaxic coordinates relative to bregma in mm: A/P −1.46, M/L ± 0.25, D/V −5.75) remain stable in the site of injection but not in the structures above the nucleus or nearby for at least 2 weeks. Notably, when the fluorescein moiety is linked to the 3′-end of the passenger strand of microRNA mimic, it does not interfere with its function. There are different moieties which could be linked to oligonucleotides to trace their biodistribution. We recommend using FAM (6-carboxyfluorescein) as despite being relatively less stable in comparison to some other dyes, it is small, cheap, exceptionally bright (the signal is ~6 times stronger than EGFP) and can be detected either by direct fluorescence analysis or by immunofluorescence (there are many well-established antibodies against FAM which represents a strong epitope) [6].

4.2 Preparation of Oligonucleotides (Refers to Section 3.2)

4.2.1. MicroRNA-mimic oligonucleotides should be ordered for in vivo use i.e. including RNase-free HPLC purification and Na^+ desalting.

4.2.2. Store lyohyphilized oligonucleotides frozen.

4.2.3. Thaw the tube on ice and spin down prior to opening.

4.2.4. Add an appropriate amount of RNAse-free sterile NaCl (up to the concentration of 10–50 mg/ml and not lower than 10 μM).

4.2.5. (refers to **step 3.2.1**) Resuspending includes:

4.2.6. pipetting,

4.2.7. vortexing,

4.2.8. spinning down,

4.2.9. keeping on ice >30 min preferably on a rocker or soft shaker to ensure complete solubilization.

4.2.10. (refers to **step 3.2.2.1**) Mix sense and antisense oligonucleotides accordingly ensuring the same volumes and the same molar concentrations of both.

4.2.11. (refers to **step 3.2.2.2**) Put the tubes to the preprogrammed PCR-cycler (preferably program 95 °C on pause before the first step, so that one can proceed immediately to the further steps) and

4.2.12. Run/unpause the program.

In case, there is no Ramp feature in the cycler, one may work-around that by programming as follows (here is a program called "Anneal-siRNA.cyc" which can be loaded per usb and used directly on Peqlab amplifiers):

```
P Anneal-siRNA.CYC        H X X 925 0 70 0
S 26.11.2010              R X X 900 X 1 X
T 21:36:36                H X X 900 0 70 0
C                         R X X 875 X 1 X
A                         H X X 875 0 70 0
U                         R X X 850 X 1 X
B 0,990,61                H X X 850 0 70 0
E X 1 1100 X X X          R X X 825 X 1 X
H X X 950 0 180 0         H X X 825 0 70 0
R X X 925 X 1 X           R X X 800 X 1 X
```

```
H X X 800 0 70 0
R X X 775 X 1 X
H X X 775 0 70 0
R X X 750 X 1 X
H X X 750 0 70 0
R X X 725 X 1 X
H X X 725 0 70 0
R X X 700 X 1 X
H X X 700 0 70 0
R X X 675 X 1 X
H X X 675 0 70 0
R X X 650 X 1 X
H X X 650 0 70 0
R X X 625 X 1 X
H X X 625 0 70 0
R X X 600 X 1 X
H X X 600 0 70 0
R X X 575 X 1 X
H X X 575 0 70 0
R X X 550 X 1 X
H X X 550 0 70 0
R X X 525 X 1 X
H X X 525 0 70 0
R X X 500 X 1 X
H X X 500 0 70 0
R X X 475 X 1 X
H X X 475 0 70 0
R X X 450 X 1 X
H X X 450 0 70 0
R X X 425 X 1 X
H X X 425 0 70 0
R X X 400 X 1 X
H X X 400 0 70 0
R X X 375 X 1 X
H X X 375 0 70 0
R X X 350 X 1 X
H X X 350 0 70 0
R X X 325 X 1 X
H X X 325 0 70 0
R X X 300 X 1 X
H X X 300 0 70 0
R X X 275 X 1 X
H X X 275 0 70 0
R X X 250 X 1 X
H X X 250 0 70 0
R X X 225 X 1 X
H X X 200 0 70 0
L X X 40 X -1 X
```

4.2.13. (refers to **steps 3.2.4** and **3.2.5**) Take the needed volume of the sense-antisense annealed solution for the chosen number of injections/infusions and mix it with the indicated/calculated amount of ACSF (depending on the intended final concentration) and HiPerfect reagent (Qiagen) according to the manufacturer's protocol and [35]. For example, mix 10 μl of the annealed oligonucleotides with 855 μl of ACSF and 135 μl HiPerfect reagent (final volume = 1000 μl) which will be sufficient to fill 5–6 100 μl-pumps, for example, model "1004" or "1002" which will perform infusion during 2 or 4 weeks, respectively. Generally, the following formula can be used to estimate the amount of liquid (V, μl) needed to fill the setup with 100 μl-pumps for several mice (n): $V = 150 \times n + 150$. The dead volume (in this case 150 μl) can be further empirically adjusted dependent on the length of the tubing, different infusion setups, volume of the pumps, etc.

Such large overheads are explained by the need to fill not only the pump which usually has a real volume of 90–100 μl but also the infusion setup.

4.2.14. (refers to **step 3.2.6**) This solution should not be frozen–thawed afterwards: use it preferably on the same day. Store this solution in the fridge and use within several hours but if possible use it directly for infusions/injections.

4.2.15. The oligonucleotides dissolved in NaCl but not mixed with ACSF/HiPerfect may be stored frozen for several months.

4.3 Assembly of the Infusion setup (Refers to Section 3.3)

4.3.1. General note: in the tested concentrations, the stability of the oligonucleotides was sufficient to withstand the long contact with RNAse-rich environment of the tissues for several weeks. Hence additional treatment of the setup (pump, tubing, flow moderator, cannulas, T-connector, etc.) with RNAse-eliminating reagents would make the procedure even more difficult but would not critically enhance the stability, and thus is not essential. However, when decreasing the concentrations of oligonucleotides it is critical to test this issue and if needed, consider eliminating even low amount of RNAses from the setup. The other possibility would be to use another design of oligonucleotides, with even more robust stability parameters [26].

4.3.2. (refers to **step 3.3.1.2**) For bilateral infusions, it is important to use two pumps connected separately to each cannula (Fig. 2a), as soon as connecting only one pump and implementing a V-connector to pump simultaneously via both cannulas will inevitably lead to different pressures or even clogging of one of the cannulas thus leading to preferential or even exclusive pumping to only one brain side. The distribution of the solution can be verified by infusing nontoxic Fast Green food dye (Fig. 2b). For further studies regarding Fast Green dye distribution using glass micropipettes, see Cunningham et al. [34]. Note the extent of destruction of the brain in Fig. 2b indicated by arrows. It is recommended to filter and spin down Fast Green dye before addition to the final infusion solution to avoid clogging the infusion system with large particles.

4.3.3. (refers to **step 3.3.2**) When the beveled pipette is fire polished, the tip ends up being more rounded or smooth. Note that polishing the beveled part at the end will make it more smooth which would dull the pipette tip. However, "polishing" the side of the shaft of the capillary can bend it. In this case the pipette would be considered "forged and bent" and not polished.

4.3.4. (refers to **step 3.3.2.1.3**) Light-sensitive glue PL 5151 may be hardened by conventional light source. Another feature is that its color is yellowish which makes it easier to visually control for example when to switch on the light source if there is a need to restrict spreading of the glue drop on a surface (as required in **steps 3.3.2.2.3.4–3.3.2.2.3.6**).

4.3.5. (refers to **step 3.3.2.1.4**) This step is advisable because increasing the area of contact between the setup and the skull predisposes that the connection between both surfaces is more tight and robust. In case of the flow moderator cap (which is not symmetric), make sure that the concavity is on the same side as the taper to make it congruent with the skull surface (*see* Figs. 4b and 7c for reference).

4.3.6. (refers to **steps 3.3.2.2**, **3.3.2.2.1.1**, and **3.3.2.2.1.2**) See the manual from Sutter Instruments and the Pipette Cookbook in the Pipette Cookbook (Sutter Instruments, http://www.sutter.com/PDFs/pipette_cookbook.pdf, p. 91) for more details about parameters. Moreover, Sutter Instruments can provide this (item number PIP50BV30) or any other type of custom-made micropipettes on demand.

If the 90 °-bent micropipettes are preferred, it is possible to bend the glass shaft on the Sutter puller while using a box or trough filament. Notably, the trough filament works better since it is easier to make a complete 90 ° bend without running into the filament.

Generally, beveled tips are safer for tissues than a blunt clean break. In case the experimenter would like to bevel the tip of either straight or 90 °-bent micropipette, one may consider using a BV-10 beveler (Sutter Instruments) with 104C plate. This step would take place after scoring and breaking and should be followed by washing of the tip (D20 rinse, and then wash with EtOH and then air or bake dry). Bevel the tip at 30 ° to create a sharp-angled tip which would be less invasive when penetrating through the brain tissue.

4.3.7. (refers to **step 3.3.2.2.2.1**) Using B100-58-15 and doing a modified one-sided manual-pull on the P-97 puller with a 2.5 or 3.0 mm box filament, breaking at the inner diameter of 50 μm will result in the final taper length of 9–11 mm.

4.3.8. (refers to **step 3.3.2.2.3.1**) If the micropipette is too long, the setup will stand out far up from the surface of the skull, which will decrease the efficiency of the wound closing. If the distance is too small, the connection between the micropipette and the T-connector might be weak, which may result in a possibility of displacement of the micropipette (Fig. 5b).

4.3.9. (refers to **step 3.3.2.2.3.2**) It is important to perform this step at the beginning in order to facilitate the manipulation with the micropipette in the next steps.

4.3.10. The easiest way to do that is to make a stand with a small hole inside. For example one can use a clean inverted eight-channel dispenser which could be easily fixed on a bench so that the eight-channel openings look upwards. Make sure that the length of the channel exceeds the distance between the taper of the micropipette and the place where it is stuck when inserted into the spacer/flow moderator cap (usually approximately 8 mm).

4.3.11. Put the spacer/flow moderator cap in the middle of one of the channels so that the hole in the spacer/flow moderator cap corresponds to the hole in the channel opening. In case of the flow moderator cap (which is not symmetric), make sure that the concavity is on the same side as the taper (*see* Figs. 5b and 7c for reference).

4.3.12. Insert the taper down through the hole. The micropipette will get thicker and will be stuck ~8 mm from the taper. If longer/smaller distances are needed, widen the hole or use spacer with a smaller hole or clog the hole partially e.g., with glue.

4.3.13. Use light-sensitive glue to stabilize the upright position. Make sure micropipette is perpendicular to the plane of the spacer/flow moderator cap.

Leave this structure as it is on the stand for the **step 3.3.2.2.3.4**.

4.3.14. (refers to **step 3.3.2.2.3.3**) Without this step, the setup will stand out far up from the surface of the skull, which will prevent the wound from closing after the operation. But if cut too close to the "trunk" of the T-connector, the connection between the micropipette and the T-connector might be weak, which may result in displacement of the micropipette (Fig. 5b).

4.3.15. (refers to **step 3.3.2.2.3.4**) In fact, one needs to join the cut "arm" with the thick end of the micropipette.

4.3.16. First apply light-sensitive glue on the outer surface of the thick end of the micropipette and the spacer/flow moderator cap. Make sure leaving the very end of cannula free of glue to prevent unintended clogging thereof.

4.3.17. Carefully place the cut "arm" of the T-connector on top of the thick end of the micropipette until it reaches the spacer and stabilize the assembly in the upright position by applying light. Make sure T-connector arms are on the same axes with micropipette and perpendicular to the plane of the spacer/flow moderator cap.

4.3.18. (refers to **step 3.3.2.2.3.5**) T-connectors should be transparent or at least semi-transparent to ensure the success of this step. Make sure the glue does not reach the inner channel of the "trunk" of the T-connector. As soon as the transparency of the T-connector is usually not complete, start to apply light slightly before it fills the "arm" as it may last 1–2 s before hardening the glue. Do not use too much glue as it will fill the "arm" too fast.

4.3.19. (refers to **step 3.3.2.2.3.6**) Make sure leaving the very end of tubing free of glue to prevent unintended clogging thereof. Apply additional glue around the "trunk" opening to ensure robust connection of both parts.

4.3.20. (refers to **step 3.3.3.1**) Avoid having bubbles in the system. It is advisable to work clean and recycle all the liquid leaking from the system during filling. See additional instructions for filling the pump in the Alzet manual.

4.4 Surgery (Refers to Section 3.4)

4.4.1. (refers to **step 3.4.6**) Consider using obvious operational instructions, always keeping in mind the concepts of aseptic and antiseptic surgery, applying panthenol crème onto the eyes of a sleeping animal, using warming pad, etc.

4.4.2. (refers to **step 3.4.6**, not needed if Plastics1 connectors are used) Use small portions of large diameter silicon tubing on both sides of the holder to assure robust contact with a T-connector.

4.4.3. (refers to **step 3.4.6**) Avoid excessive tightening the screw of the holder because then it will be problematic to unscrew after the implantation without displacing the fixed mouse with implanted cannula, the skull may be damaged, the cannula may be displaced or pulled out.

4.4.4. (refers to **step 3.4.7**, not needed if the setup is not equipped with infusion kit3 spacer or flow moderator cap) Due to a large surface of the flow moderator cap or infusion kit3 spacer, it is usually impossible to see the taper and the skull under the taper when adjusting the stereotaxic coordinates. For such cases, a movable microscope arm is preferential, so that observation from the side is possible. If a commercially distributable microscope system for this purpose is not available, the light microscopes can usually be reconstructed to assure such functionality.

4.4.5. (refers to **step 3.4.8**) Small elongated paper stripes can be prepared from wedged filter paper sponge bars and put under the skin surrounding the holes. This (1) keeps the surface dry and (2) protects the mouse tissues from unintended smearing by glue. Contamination with glue is extremely undesirable, as it may damage the skin so the wound cannot be properly closed. Hence, the setup will be exposed after operation which then may lead to displacement of the cannula.

4.4.6. (refers to **steps 3.4.9** and **3.4.11**) Putting too much gel-consistency glue will increase the waiting time. Try to avoid letting the operation + waiting times to last too long (30–40 min per animal or less should be the goal) in order to reduce the stress load on the animal.

4.4.7. (refers to **step 3.4.12**) This ensures very slow opening of the holder and prevents occasional vibrations/displacements, etc.

4.4.8. (refers to **steps 3.4.13** and **3.4.14**) The arm may be cut by scissors, but preferably use cut-off disc mini drill. It is important to release the mouse from the frame prior to that to avoid additional tension on the skull during the removal of the upper arm.

4.4.9. (refers to **step 3.4.15**) Especially important for the Plastics1 connector setup is rehydrating the mouse. Inject up to 0.5 ml sterile saline s.c. and/or up to 0.5 ml i.p. to assist faster recovery.

4.4.10. (refers to **step 3.4.15**) Make sure closing the wound completely so that no parts of the setup are exposed from under the skin.

4.4.11. After the period of infusion, e.g., not later than 28 days after the operation when Model "1004" pumps were used, if no further experiments are planned, sacrifice the mice and remove the pumps. Measure the amount of the remaining liquid in the pumps to ensure proper infusion.

4.4.12. Check for damage of the brain structures (Fig. **6**) for example by Nissl staining [6] to exclude the possibility that the observed phenotype is caused by disruption of the tissues.

Acknowledgments

The authors declare no competing financial interests. This work has been supported by the DFG through SFB488, the EU through grant LSHM-CT-2005-018652 (CRESCENDO), the BMBF through NGFNplus grants FZK-01GS08153 and 01GS08142, and the HGF through Initiative CoReNe (Network II, E2) the National Science Centre (Poland) grant (SONATA) 2011/01/D/NZ4/03744, grant (HARMONIA) 2013/08/M/NZ3/01045, and the Academy of Finland. We thank Jörg Krummheuer for the protocol for preparation of the LNA-oligonucleotides for injection as well as assistance in developing the oligonucleotide synthesis strategy, Günther Schütz for support, Lena Roth for assistance with the infusion technique. Special thanks goes to Adair Oesterle for assistance with techniques to fabricate micropipettes.

References

1. Bartel DP (2009) MicroRNAs: target recognition and regulatory functions. Cell 136 (2):215–233. doi:10.1016/j.cell.2009.01.002, S0092-8674(09)00008-7 [pii]
2. Mikl M, Vendra G, Doyle M, Kiebler MA (2010) RNA localization in neurite morphogenesis and synaptic regulation: current evidence and novel approaches. J Comp Physiol A Neuroethol Sens Neural Behav Physiol 196 (5):321–334. doi:10.1007/s00359-010-0520-x
3. Schratt G (2009) microRNAs at the synapse. Nat Rev Neurosci 10(12):842–849. doi:10.1038/nrn2763, nrn2763 [pii]
4. Konopka W, Schutz G, Kaczmarek L (2011) The microRNA contribution to learning and memory. Neuroscientist 17(5):468–474. doi:10.1177/1073858411411721
5. Konopka W, Kiryk A, Novak M, Herwerth M, Parkitna JR, Wawrzyniak M, Kowarsch A, Michaluk P, Dzwonek J, Arnsperger T, Wilczynski G, Merkenschlager M, Theis FJ, Kohr G, Kaczmarek L, Schutz G (2010) MicroRNA loss enhances learning and memory in mice. J Neurosci 30(44):14835–14842. doi:10.1523/JNEUROSCI.3030-10.2010, 30/44/14835 [pii]
6. Vinnikov IA, Hajdukiewicz K, Reymann J, Beneke J, Czajkowski R, Roth LC, Novak M, Roller A, Dörner N, Starkuviene V, Theis FJ, Erfle H, Schütz G, Grinevich V, Konopka W (2014) Hypothalamic miR-103 protects from hyperphagic obesity in mice. J Neurosci 34 (32):10659–10674. doi:10.1523/jneurosci.4251-13.2014
7. Verma P, Augustine GJ, Ammar MR, Tashiro A, Cohen SM (2015) A neuroprotective role for microRNA miR-1000 mediated by limiting glutamate excitotoxicity. Nat Neurosci 18 (3):379–385. doi:10.1038/nn.3935
8. Gao J, Wang WY, Mao YW, Graff J, Guan JS, Pan L, Mak G, Kim D, Su SC, Tsai LH (2010) A novel pathway regulates memory and plasticity via SIRT1 and miR-134. Nature 466 (7310):1105–1109. doi:10.1038/nature09271, nature09271 [pii]
9. Tan CL, Plotkin JL, Veno MT, von Schimmelmann M, Feinberg P, Mann S, Handler A, Kjems J, Surmeier DJ, O'Carroll D, Greengard P, Schaefer A (2013) MicroRNA-128 governs neuronal excitability and motor behavior in mice. Science 342(6163):1254–1258. doi:10.1126/science.1244193
10. Xie J, Ameres SL, Friedline R, Hung JH, Zhang Y, Xie Q, Zhong L, Su Q, He R, Li M, Li H, Mu X, Zhang H, Broderick JA, Kim JK, Weng Z, Flotte TR, Zamore PD, Gao G (2012) Long-term, efficient inhibition of microRNA function in mice using rAAV vectors. Nat Methods 9(4):403–409. doi:10.1038/nmeth.1903
11. Ran FA, Cong L, Yan WX, Scott DA, Gootenberg JS, Kriz AJ, Zetsche B, Shalem O, Wu X, Makarova KS, Koonin EV, Sharp PA, Zhang F (2015) In vivo genome editing using Staphylococcus aureus Cas9. Nature. doi:10.1038/nature14299
12. Platt RJ, Chen S, Zhou Y, Yim MJ, Swiech L, Kempton HR, Dahlman JE, Parnas O, Eisenhaure TM, Jovanovic M, Graham DB, Jhunjhunwala S, Heidenreich M, Xavier RJ, Langer R, Anderson DG, Hacohen N, Regev A, Feng G, Sharp PA, Zhang F (2014) CRISPR-Cas9 knockin mice for genome editing and cancer modeling. Cell 159(2):440–455. doi:10.1016/j.cell.2014.09.014
13. Christensen M, Larsen LA, Kauppinen S, Schratt G (2010) Recombinant adeno-associated virus-mediated microRNA delivery into the postnatal mouse brain reveals a role for miR-134 in dendritogenesis in vivo. Front Neural Circuits 3:16. doi:10.3389/neuro.04.016.2009
14. Zovoilis A, Agbemenyah HY, Agis-Balboa RC, Stilling RM, Edbauer D, Rao P, Farinelli L, Delalle I, Schmitt A, Falkai P, Bahari-Javan S, Burkhardt S, Sananbenesi F, Fischer A (2011) microRNA-34c is a novel target to treat dementias. EMBO J 30(20):4299–4308. doi:10.1038/emboj.2011.327
15. Krützfeldt J, Rajewsky N, Braich R, Rajeev KG, Tuschl T, Manoharan M, Stoffel M (2005)

Silencing of microRNAs in vivo with 'antagomirs'. Nature 438(7068):685–689. doi:10.1038/nature04303

16. Krutzfeldt J, Kuwajima S, Braich R, Rajeev KG, Pena J, Tuschl T, Manoharan M, Stoffel M (2007) Specificity, duplex degradation and subcellular localization of antagomirs. Nucleic Acids Res 35(9):2885–2892. doi:10.1093/nar/gkm024, gkm024 [pii]
17. Esau C, Davis S, Murray SF, Yu XX, Pandey SK, Pear M, Watts L, Booten SL, Graham M, McKay R, Subramaniam A, Propp S, Lollo BA, Freier S, Bennett CF, Bhanot S, Monia BP (2006) miR-122 regulation of lipid metabolism revealed by in vivo antisense targeting. Cell Metab 3(2):87–98. doi:10.1016/j.cmet.2006.01.005
18. Liu J, Yu D, Aiba Y, Pendergraff H, Swayze EE, Lima WF, Hu J, Prakash TP, Corey DR (2013) ss-siRNAs allele selectively inhibit ataxin-3 expression: multiple mechanisms for an alternative gene silencing strategy. Nucleic Acids Res 41(20):9570–9583. doi:10.1093/nar/gkt693
19. Gatfield D, Le Martelot G, Vejnar CE, Gerlach D, Schaad O, Fleury-Olela F, Ruskeepaa AL, Oresic M, Esau CC, Zdobnov EM, Schibler U (2009) Integration of microRNA miR-122 in hepatic circadian gene expression. Genes Dev 23(11):1313–1326. doi:10.1101/gad.1781009
20. Rayner KJ, Esau CC, Hussain FN, McDaniel AL, Marshall SM, van Gils JM, Ray TD, Sheedy FJ, Goedeke L, Liu X, Khatsenko OG, Kaimal V, Lees CJ, Fernandez-Hernando C, Fisher EA, Temel RE, Moore KJ (2011) Inhibition of miR-33a/b in non-human primates raises plasma HDL and lowers VLDL triglycerides. Nature 478(7369):404–407. doi:10.1038/nature10486
21. Elmen J, Lindow M, Silahtaroglu A, Bak M, Christensen M, Lind-Thomsen A, Hedtjarn M, Hansen JB, Hansen HF, Straarup EM, McCullagh K, Kearney P, Kauppinen S (2007) Antagonism of microRNA-122 in mice by systemically administered LNA-antimiR leads to up-regulation of a large set of predicted target mRNAs in the liver. Nucleic Acids Res 36(4):1153–1162. doi:10.1093/nar/gkm1113
22. Elmén J, Lindow M, Schütz S, Lawrence M, Petri A, Obad S, Lindholm M, Hedtjärn M, Hansen HF, Berger U, Gullans S, Kearney P, Sarnow P, Straarup EM, Kauppinen S (2008) LNA-mediated microRNA silencing in non-human primates. Nature 452(7189):896–899. doi:10.1038/nature06783
23. Lanford RE, Hildebrandt-Eriksen ES, Petri A, Persson R, Lindow M, Munk ME, Kauppinen S, Orum H (2009) Therapeutic silencing of microRNA-122 in primates with chronic hepatitis C virus infection. Science 327(5962):198–201. doi:10.1126/science.1178178
24. Zhang Y, Wang Z, Gemeinhart RA (2013) Progress in microRNA delivery. J Control Release 172(3):962–974. doi:10.1016/j.jconrel.2013.09.015
25. Elmen J, Thonberg H, Ljungberg K, Frieden M, Westergaard M, Xu Y, Wahren B, Liang Z, Orum H, Koch T, Wahlestedt C (2005) Locked nucleic acid (LNA) mediated improvements in siRNA stability and functionality. Nucleic Acids Res 33(1):439–447. doi:10.1093/nar/gki193, 33/1/439 [pii]
26. Yu D, Pendergraff H, Liu J, Kordasiewicz HB, Cleveland DW, Swayze EE, Lima WF, Crooke ST, Prakash TP, Corey DR (2012) Single-stranded RNAs use RNAi to potently and allele-selectively inhibit mutant huntingtin expression. Cell 150(5):895–908. doi:10.1016/j.cell.2012.08.002
27. Mook OR, Baas F, de Wissel MB, Fluiter K (2007) Evaluation of locked nucleic acid-modified small interfering RNA in vitro and in vivo. Mol Cancer Ther 6(3):833–843. doi:10.1158/1535-7163.mct-06-0195
28. Mong JA, Devidze N, Goodwillie A, Pfaff DW (2003) Reduction of lipocalin-type prostaglandin D synthase in the preoptic area of female mice mimics estradiol effects on arousal and sex behavior. Proc Natl Acad Sci U S A 100(25):15206–15211. doi:10.1073/pnas.2436540100
29. Bramsen JB, Laursen MB, Damgaard CK, Lena SW, Ravindra Babu B, Wengel J, Kjems J (2007) Improved silencing properties using small internally segmented interfering RNAs. Nucleic Acids Res 35(17):5886–5897. doi:10.1093/nar/gkm548
30. Su J, Baigude H, McCarroll J, Rana TM (2011) Silencing microRNA by interfering nanoparticles in mice. Nucleic Acids Res 39(6):e38. doi:10.1093/nar/gkq1307
31. John M, Constien R, Akinc A, Goldberg M, Moon Y-A, Spranger M, Hadwiger P, Soutschek J, Vornlocher H-P, Manoharan M, Stoffel M, Langer R, Anderson DG, Horton JD, Koteliansky V, Bumcrot D (2007) Effective RNAi-mediated gene silencing without interruption of the endogenous microRNA pathway. Nature 449(7163):745–747. doi:10.1038/nature06179

32. Trajkovski M, Hausser J, Soutschek J, Bhat B, Akin A, Zavolan M, Heim MH, Stoffel M (2011) MicroRNAs 103 and 107 regulate insulin sensitivity. Nature 474 (7353):649–653. doi:10.1038/nature10112, nature10112 [pii]

33. Lee ST, Chu K, Jung KH, Kim JH, Huh JY, Yoon H, Park DK, Lim JY, Kim JM, Jeon D, Ryu H, Lee SK, Kim M, Roh JK (2012) miR-206 regulates brain-derived neurotrophic factor in Alzheimer disease model. Ann Neurol 72 (2):269–277. doi:10.1002/ana.23588

34. Cunningham MG, O'Connor RP, Wong SE (2008) Construction and implantation of a microinfusion system for sustained delivery of neuroactive agents. J Vis Exp. doi:10.3791/716, 716 [pii]

35. Zovoilis A, Agbemenyah HY, Agis-Balboa RC, Stilling RM, Edbauer D, Rao P, Farinelli L, Delalle I, Schmitt A, Falkai P, Bahari-Javan S, Burkhardt S, Sananbenesi F, Fischer A (2011) microRNA-34c is a novel target to treat dementias. EMBO J 30(20):4299–4308. doi:10.1038/emboj.2011.327, emboj2011327 [pii]

36. Paxinos G, Franklin KBJ (2001) The mouse brain in stereotaxic coordinates, 2nd edn. Academic Press, San Diego

Neuromethods (2017) 128: 119–127
DOI 10.1007/7657_2016_4

Published online: 6 August 2016

Isolating and Screening Subcellular miRNAs in Neuron

Min Jeong Kye

Abstract

Since local protein synthesis is proposed to explain spatial and temporal regulation of gene expression in highly polarized cells such as neurons, various mechanisms for regulating protein synthesis are suggested. Among them, microRNA (miRNA) is one of the key regulators for protein synthesis in the synaptic area. As miRNAs can selectively repress mRNA translation with sequence-specific manner, profiling of miRNAs located in synaptic area has been an important topic to understand the function of neuronal miRNAs. Interestingly, many miRNAs are detected in the synaptic area, but their subcellular distribution in neuron varies. This suggests that there are cellular mechanisms actively regulating miRNA expression and localization in subcellular compartments of neurons. In this chapter, we review currently available methods of synaptic miRNA profiling from isolating samples, purifying RNAs, and measuring expression.

Keywords: MicroRNA, Synaptic compartment, Laser capture microdissection, Boyden chamber, Synaptosome, Neuron

1 Introduction

As neurons are highly polarized and neuronal function often requires immediate responses from external stimuli, spatial and temporal control of protein synthesis is crucial for properly operating nervous system. To explain this phenomenon, the idea of subcellular regulation of protein synthesis has been suggested and various evidences support this [1–3]. MicroRNAs are a subclass of small noncoding RNA and their most well-known function is repression of mRNA translation [4]. Proteins composed of functional complex for microRNA, RNA-induced silencing complex (RISC), are detected in synaptic area such as growth cones and post-synaptic densities, and one of the RISC components, MOV10, responds to neuronal activity at the post-synaptic sites [5–7]. Moreover, other proteins important for microRNA function such as FMRP, DICER, and elF2c are also present at the synaptic site [5, 6]. Importantly, numerous microRNAs are detected from synaptic area including pre- and post-synaptic compartments of various neurons together with their target mRNAs [8, 9]. Modulating expression of those microRNAs changes animal behavior requiring synaptic function such as learning and memory formation and neuromuscular activities [10, 11].

MicroRNAs are expressed in most of the tissues and cells, and their dysregulation often causes diseases such as cancers, metabolic diseases, and neurodegenerative disorders [12–14]. Dysregulated miRNA expression is reported in many different neurological disorders such as Alzheimer's disease (AD), Parkinson's disease (PD), amyotrophic lateral sclerosis (ALS), and spinal muscular atrophy (SMA) [15–20]. As synapse is often considered as a primary site showing defects for the neurological disorders such as ALS, SMA, and AD, it is very important to understand regulation of gene expression at the synapse [21–24].

Here, we summarize currently available methods of screening miRNAs from subcellular compartment of neurons.

2 Materials

2.1 Primary Neuron Culture

Embryonic mouse (E13.5 for motor neuron, E17 for hippocampal and cortical neurons).

Poly-D-lysine (Sigma-Aldrich).

Trypsin (Worthington).

DNase I (Worthington).

Neurobasal media (Thermo Fisher Scientific).

Minimum Essential Media (Thermo Fisher Scientific).

Glucose (Sigma-Aldrich).

Fetal Calf Serum (Biochrome).

B27 (Thermo Fisher Scientific).

Pen/Strep (Thermo Fisher Scientific).

Amphotericin B (Thermo Fisher Scientific).

PBS (Thermo Fisher Scientific).

Cytosine arabinoside (Sigma-Aldrich).

Tweezers.

Scissors.

Dissecting microscope.

2.2 Laser Capture Microdissection

PixCell® IIe LCM System (Arcturus Laser capture microdissection microscope, currently Thermo Fisher Scientific).

Arcturus Laser capture microdissection system (Thermo Fisher Scientific).

Glass cover slip.

Poly-D-lysine (Sigma-Aldrich).

Xylene.

Ethanol.

2.3 Modified Boyden Chamber

Transwell (3 μm pore, Permeable Support, Corning).

Brain-derived neurotrophic factor (BDNF, Peprotech).

Glial cell line-derived neurotrophic factor (GCNF, Peprotech).

Ciliary neurotrophic factor (CNTF, Peprotech).

Poly-D-lysine (Sigma-Aldrich).

2.4 Synaptoneurosome Preparation

Synaptosome preparation kit (SynPER Synaptic Protein Extraction Reagent, Thermo Fisher Scientific).

2.5 Isolation of Total RNA

mirVana total RNA isolation kit (Thermo Fisher Scientific).

Trizol (Thermo Fisher Scientific).

Trizol LS Reagent (Thermo Fisher Scientific).

2.6 Taqman-Based Real-Time PCR to Measure miRNAs

Real-time PCR machine.

Taqman MicroRNA assays and arrays (Thermo Fischer Scientific).

High Capacity cDNA Archive kit (Thermo Fisher Scientific).

2.7 Additional Reagent and Instrument

Quant-iT RiboGreen (Thermo Fisher Scientific).

Agilent Bioanalyzer (Thermo Fisher Scientific).

Nanodrop (Thermo Fisher Scientific).

3 Methods

3.1 Laser Capture Microdissection Methods

3.1.1 Hippocampal Neuron Culture for Laser Capture Experiment

(a) Hippocampi were dissected from E18 mouse embryos.

(b) Neurons were dissociated with trypsin (0.01 % for 5 min at RT), triturated with DNase I (0.1 %) in plating media (minimum essential medium, 0.6 % glucose, 5 % FBS, and 1× pen-strep), and plated on poly-D-lysine-coated plates.

(c) Cells were plated at 60,000 cells/cover slip for laser capture microdissection experiment. Plating medium was changed after 4 h and cells were maintained in neurobasal medium, 2 % B27 supplement, 500 μM L-glutamine, and 1× pen-strep at 37 °C in a humidified incubator with 5 % CO_2.

(d) After 4–5 days in culture, one-half of the media was changed and cytosine arabinoside (Sigma) was added to a final concentration of 1 μM.

3.1.2 Cell Staining and Laser Capture Microdissection

(a) Hippocampal cultures were washed with PBS and fixed in ice-cold 70 % ethanol. To avoid RNA degradation, staining was performed with solutions prepared with DEPC-treated water on ice block.

(b) Cells were sequentially dehydrated in 95 and 100 % ethanol and sequentially rehydrated in 95 % ethanol, 70 % ethanol, and TBST (0.05 M Tris base, 0.9 % NaCl, 0.1 % Tween-20).

(c) Cells were stained for 3 min in 1:100 anti-GFAP-Cy3-conjugated monoclonal antibody (Sigma) diluted in TBST; washed three times in TBST and three times in TBS (0.05 M Tris base, 0.9 % NaCl); and sequentially dehydrated in 70, 95, and 100 % ethanol, followed by a rinse in xylenes at RT.

(d) Slides were air-dried, and neurites and neuronal cell bodies were captured using a PixCell® IIe LCM System (Thermo Fisher Scientific). Glial cells were identified by GFAP-positive staining and neuronal neurites were captured from regions without GFAP staining (Fig. 1).

3.1.3 Isolating RNA

Captured fractions are buffer exchanged into water using Centri-Spin 10 columns (Princeton Separations). Samples are treated with 1 μL of RNase inhibitors (Eppendorf) and stored in −80 °C for further analysis.

3.2 Modified Boyden Chamber

Second method to isolate RNAs from neurite compartment is using modified Boyden chamber. This is more efficient and less time consuming than laser capture microdissection. However, it can contain contaminants such as glia and part of somata.

3.2.1 Cortical Neuron Culture for Modified Boyden Chamber Experiment

(a) Cortices were dissected from E18 mouse embryos.

(b) Neurons were mechanically triturated with pipettes and 1000,000 cells/well were plated on poly-D-lysine-coated membranes of Transwell (3 μm pore membrane, Corning).

(c) Cells were maintained in neurobasal medium, 2 % B27 supplement, 500 μM L-glutamine, and 1× pen-strep at 37 °C in a humidified incubator with 5% CO_2. Neurotrophic growth factors (100 ng/mL BDNF, 100 ng/mL GDNF, and 50 ng/mL GCNF) were added to lower part of chamber.

(d) Every 4–5 days in culture, one-half of the media was changed from both sides of chambers (Fig. 2).

3.2.2 Spinal Motor Neuron Culture for Modified Boyden Chamber Experiment

(a) Spinal cords were dissected from E13.5 mouse embryos.

(b) Cells were dissociated with trypsin (0.01 % for 5 min at RT), triturated with DNase I (0.1 %) in plating media (minimum essential medium, 0.6 % glucose, 5 % FBS, and 1× pen-strep), and plated on poly-D-lysine-coated Transwell.

(c) Cells were maintained in neurobasal medium, 2 % B27 supplement, 500 μM L-glutamine, and 1× pen-strep at 37 °C in a humidified incubator with 5 % CO_2. Neurotrophic growth

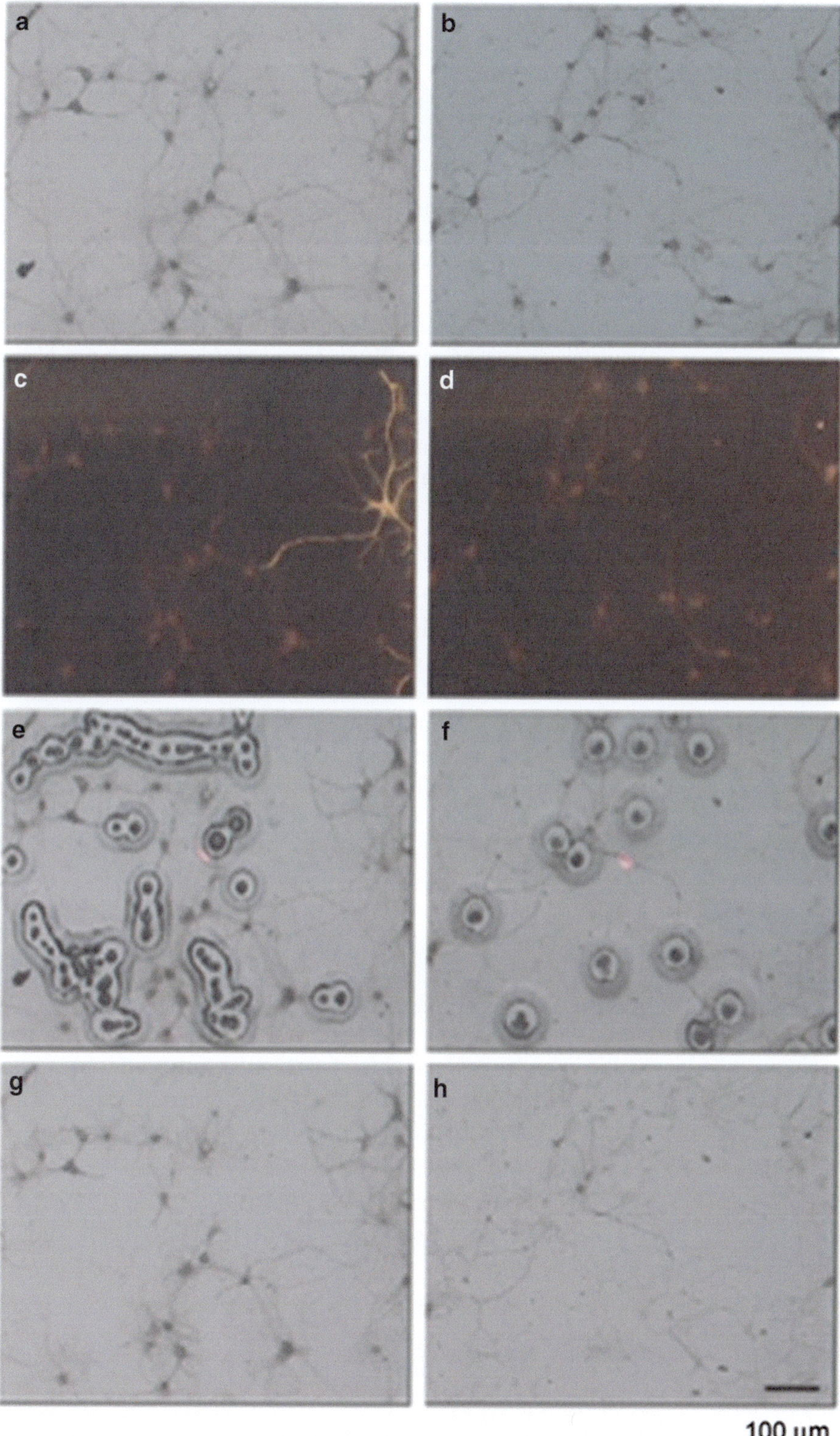

Fig. 1 Laser capture microdissection of neurites. Bright-field and fluorescent microscopy was used to isolate glial cell-free neurites (**a**, **c**, **e**, **g**) and cell body (**b**, **d**, **f**, **h**). Representative images: neuron cultures before laser capture with bright field (**a**, **b**), fluorescence with GFAP staining to mark glial cells (**c**, **d**), laser capture (**e**, **f**), and culture slides after laser capture with bright field (**g**, **h**) (images are taken from [8])

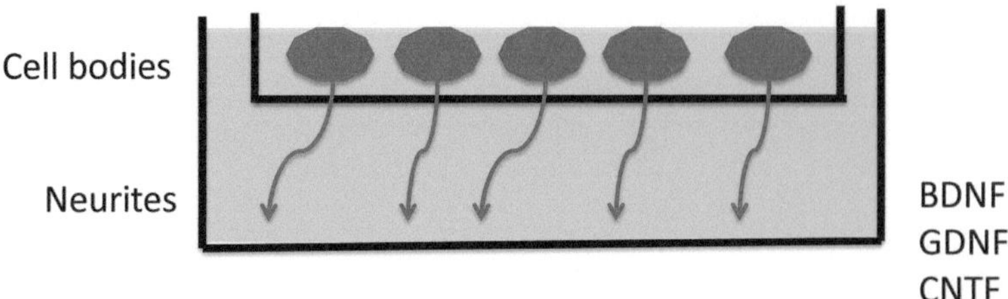

Fig. 2 Schematic drawing of modified Boyden chamber to isolate neurite compartments from neurons. Neurons were seeded on upper side of membrane, which has 3 μm of pore. Lower chamber contains growth factors (100 ng/μL GDNF, 100 ng/μL BDNF, and 50 ng/μL CNTF) to promote survival and growth of neurons. After 10 days in culture, neurite compartment can be isolated from bottom side of membrane

factors (100 ng/mL BDNF, 100 ng/mL GDNF, and 50 ng/mL CNTF) were added to lower part of chamber.

(d) Every 4–5 days in culture, one-half of the media was changed from both sides of chambers.

3.2.3 Isolating Total RNA from Samples

After 10 days of culture, samples were collected using scrapper from upper side of the membrane (containing mainly cell bodies, proximal axons, and dendrites) and lower side of the membrane (containing mainly distal axons and some distal dendrites). Total RNA was extracted from those samples using *mirVana* total RNA isolation kit (Thermo Fisher Scientific).

3.3 Isolation of Synaptosome

The third method to separate subcellular compartments in neurons is isolating synaptosome. This method can be used for cultured neurons as well as brain tissues.

3.3.1 From Cultured Neurons

(a) For cultured neuron, 14–21-day cultured neurons can be used. Cell medium is removed and cells are washed with PBS. Syn-PER synaptic protein extraction reagent is added and samples are collected with cell scrapper.

(b) Samples are centrifuged with 1200 × *g* for 10 min at 4 °C to separate nucleus (pellet) from the cytoplasm and neurites (supernatant).

(c) Collected supernatant is again centrifuged with 15,000 × *g* for 20 min at 4 °C. In this step, we separate synaptic compartment (pellet) from soma (supernatant).

(d) Total RNA can be isolated from cellular fractions using Trizol (for pellet) or Trizol LS reagent (supernatant).

3.3.2 From Brain Samples

(a) For brain tissue, 10 mL of Syn-PER synaptic protein extraction reagent is added per 1 g of brain tissue, and brain tissue is homogenized using Dounce tissue grinder on ice.

Alternatively, syringe with needles can be used to homogenize brain tissues. Needles need to be changed serially from big ones to small ones to achieve proper dissociation.

(b) After homogenization, samples are centrifuged with 1200 × *g* for 10 min at 4 °C to separated nucleus (pellet) from the cytoplasm and neurites (supernatant). Supernatant needs to be transferred to new tube.

(c) Collected supernatant is again centrifuged with 15,000 × *g* for 20 min at 4 °C. In this step, we separate synaptic compartment (pellet) from soma (supernatant).

(d) Total RNA can be isolated from cellular fractions using Trizol (for pellet) or Trizol LS reagent (supernatant).

3.4 Measuring miRNA Expression Using Taqman-Based Real-Time PCR

From the isolated samples, total RNA is extracted and quantified. Total RNA can be isolated using Trizol or *mirVana* total RNA isolation kit. However, it is important to choose right RNA extraction method to obtain good quality and enough amount of RNA for further analysis. When samples are plenty, for example, synaptosomes from brain tissues, Trizol works well. However, when the amount of sample is limited such as modified Boyden chamber method or synaptosomes from cultured neurons, *mirVana* total RNA isolation kit is preferred. For the laser capture microdissection method, we only used buffer exchange kit to minimize the additional loss of RNAs.

After extracting total RNAs from samples, the amount of RNA can be quantified using Nanodrop or Quant-iT RiboGreen RNA assay kit. The quality and integrity of RNA can be measured with Agilent Bioanalyzer (Thermo Fisher Scientific).

To measure miRNA amount in samples, Taqman-based real-time PCR is commonly used. For this, we produce cDNA from total RNA using High Capacity cDNA Archive kit (Thermo Fisher Scientific). We use miRNA-specific primers to produce individual cDNA for miRNA and quantify miRNA amount using Taqman-based real-time PCR (Taqman-miRNA assay, Thermo Fisher Scientific). Furthermore, those RNAs can be used for genome-wide screening such as next generation sequencing and microarray.

4 Conclusions

Various methods to isolate synaptic compartment from neuron have been developed to characterize the composition of the synaptic area. As synaptic compartments contain rather different composition of molecules compared to soma, it has been always interesting to know how molecules are sorted, trafficked, produced, destroyed, and recycled in synaptic compartments. MicroRNA-mediated translational repression is one of the proposed mechanisms to regulate synaptic gene expression. To understand this mechanism, knowing which miRNAs function in the synaptic area will be very helpful. Various scientific reports listed miRNAs in the synaptic area of different neuronal subtypes, but those lists are not yet complete. For example, numbers of miRNAs were dysregulaed in synaptic areas of neurons with genetic disorders, prion disease, and alcohol abuse [20, 25, 26].

It seems that "perfect" method for isolating synaptic compartment of neurons is not yet available. Therefore, we can combine two or more methods to support weak point of each method. Here we summarized the most commonly used three methods to isolate synaptic compartments from neurons. In addition, we shortly discussed how to measure miRNA expression.

Acknowledgement

This work is supported by Deutsche Forschungsgemeinschaft (German Research Foundation), University of Cologne (Cologne Fortune), and cure SMA.

References

1. Holt CE, Schuman EM (2013) The central dogma decentralized: new perspectives on RNA function and local translation in neurons. Neuron 80(3):648–657. doi:10.1016/j.neuron.2013.10.036, S0896-6273(13)00988-4 [pii]
2. Sutton MA, Schuman EM (2006) Dendritic protein synthesis, synaptic plasticity, and memory. Cell 127(1):49–58. doi:10.1016/j.cell.2006.09.014, S0092-8674(06)01206-2 [pii]
3. Jung H, Yoon BC, Holt CE (2012) Axonal mRNA localization and local protein synthesis in nervous system assembly, maintenance and repair. Nat Rev Neurosci 13(5):308–324. doi:10.1038/nrn3210, nrn3210 [pii]
4. Kosik KS (2006) The neuronal microRNA system. Nat Rev Neurosci 7(12):911–920. doi:10.1038/nrn2037, nrn2037 [pii]
5. Hengst U, Cox LJ, Macosko EZ, Jaffrey SR (2006) Functional and selective RNA interference in developing axons and growth cones. J Neurosci 26(21):5727–5732. doi:10.1523/JNEUROSCI.5229-05.2006, 26/21/5727 [pii]
6. Lugli G, Torvik VI, Larson J, Smalheiser NR (2008) Expression of microRNAs and their precursors in synaptic fractions of adult mouse forebrain. J Neurochem 106(2):650–661. doi:10.1111/j.1471-4159.2008.05413.x, JNC5413 [pii]
7. Banerjee S, Neveu P, Kosik KS (2009) A coordinated local translational control point at the synapse involving relief from silencing and MOV10 degradation. Neuron 64(6):871–884. doi:10.1016/j.neuron.2009.11.023, S0896-6273(09)00939-8 [pii]
8. Kye MJ, Liu T, Levy SF, Xu NL, Groves BB, Bonneau R, Lao K, Kosik KS (2007) Somatodendritic microRNAs identified by laser capture and multiplex RT-PCR. RNA 13(8):1224–1234. doi:10.1261/rna.480407, rna.480407 [pii]
9. Natera-Naranjo O, Aschrafi A, Gioio AE, Kaplan BB (2010) Identification and quantitative analyses of microRNAs located in the distal axons of sympathetic neurons. RNA 16(8):1516–1529. doi:10.1261/rna.1833310, rna.1833310 [pii]
10. Gao J, Wang WY, Mao YW, Graff J, Guan JS, Pan L, Mak G, Kim D, Su SC, Tsai LH (2010) A novel pathway regulates memory and plasticity via SIRT1 and miR-134. Nature 466(7310):1105–1109. doi:10.1038/nature09271, nature09271 [pii]
11. Amin ND, Bai G, Klug JR, Bonanomi D, Pankratz MT, Gifford WD, Hinckley CA, Sternfeld MJ, Driscoll SP, Dominguez B, Lee KF, Jin X, Pfaff SL (2015) Loss of motoneuron-specific microRNA-218 causes systemic neuromuscular failure. Science 350(6267):1525–1529.

doi:10.1126/science.aad2509, 350/6267/1525 [pii]

12. Medina PP, Nolde M, Slack FJ (2010) OncomiR addiction in an in vivo model of microRNA-21-induced pre-B-cell lymphoma. Nature 467(7311):86–90. doi:10.1038/nature09284, nature09284 [pii]
13. Vienberg S, Geiger J, Madsen S, Dalgaard LT (2016) MicroRNAs in metabolism. Acta Physiol (Oxf). doi:10.1111/apha.12681
14. Kye MJ, Goncalves Ido C (2014) The role of miRNA in motor neuron disease. Front Cell Neurosci 8:15. doi:10.3389/fncel.2014.00015
15. Hebert SS, Horre K, Nicolai L, Papadopoulou AS, Mandemakers W, Silahtaroglu AN, Kauppinen S, Delacourte A, De Strooper B (2008) Loss of microRNA cluster miR-29a/b-1 in sporadic Alzheimer's disease correlates with increased BACE1/beta-secretase expression. Proc Natl Acad Sci U S A 105 (17):6415–6420. doi:10.1073/pnas.0710263105, 0710263105 [pii]
16. Wang WX, Rajeev BW, Stromberg AJ, Ren N, Tang G, Huang Q, Rigoutsos I, Nelson PT (2008) The expression of microRNA miR-107 decreases early in Alzheimer's disease and may accelerate disease progression through regulation of beta-site amyloid precursor protein-cleaving enzyme 1. J Neurosci 28 (5):1213–1223. doi:10.1523/JNEUROSCI.5065-07.2008, 28/5/1213 [pii]
17. Gui Y, Liu H, Zhang L, Lv W, Hu X (2015) Altered microRNA profiles in cerebrospinal fluid exosome in Parkinson disease and Alzheimer disease. Oncotarget 6(35):37043–37053. doi:10.18632/oncotarget.6158, 6158 [pii]
18. Briggs CE, Wang Y, Kong B, Woo TU, Iyer LK, Sonntag KC (2015) Midbrain dopamine neurons in Parkinson's disease exhibit a dysregulated miRNA and target-gene network. Brain Res 1618:111–121. doi:10.1016/j.brainres.2015.05.021, S0006-8993(15)00423-0 [pii]
19. Williams AH, Valdez G, Moresi V, Qi X, McAnally J, Elliott JL, Bassel-Duby R, Sanes JR, Olson EN (2009) MicroRNA-206 delays ALS progression and promotes regeneration of neuromuscular synapses in mice. Science 326 (5959):1549–1554. doi:10.1126/science.1181046, 326/5959/1549 [pii]
20. Kye MJ, Niederst ED, Wertz MH, Goncalves ID, Akten B, Dover KZ, Peters M, Riessland M, Neveu P, Wirth B, Kosik KS, Sardi SP, Monani UR, Passini MA, Sahin M (2014) SMN regulates axonal local translation via miR-183/mTOR pathway. Hum Mol Genet. doi:10.1093/hmg/ddu350, ddu350 [pii]
21. Boido M, Vercelli A (2016) Neuromuscular junctions as key contributors and therapeutic targets in spinal muscular atrophy. Front Neuroanat 10:6. doi:10.3389/fnana.2016.00006
22. Armstrong GA, Drapeau P (2013) Loss and gain of FUS function impair neuromuscular synaptic transmission in a genetic model of ALS. Hum Mol Genet 22(21):4282–4292. doi:10.1093/hmg/ddt278, ddt278 [pii]
23. Machamer JB, Collins SE, Lloyd TE (2014) The ALS gene FUS regulates synaptic transmission at the Drosophila neuromuscular junction. Hum Mol Genet 23(14):3810–3822. doi:10.1093/hmg/ddu094, ddu094 [pii]
24. Di J, Cohen LS, Corbo CP, Phillips GR, El Idrissi A, Alonso AD (2016) Abnormal tau induces cognitive impairment through two different mechanisms: synaptic dysfunction and neuronal loss. Sci Rep 6:20833. doi:10.1038/srep20833, srep20833 [pii]
25. Boese AS, Saba R, Campbell K, Majer A, Medina S, Burton L, Booth TF, Chong P, Westmacott G, Dutta SM, Saba JA, Booth SA (2016) MicroRNA abundance is altered in synaptoneurosomes during prion disease. Mol Cell Neurosci 71:13–24. doi:10.1016/j.mcn.2015.12.001, S1044-7431(15)30041-5 [pii]
26. Most D, Leiter C, Blednov YA, Harris RA, Mayfield RD (2016) Synaptic microRNAs coordinately regulate synaptic mRNAs: perturbation by chronic alcohol consumption. Neuropsychopharmacology 41(2):538–548. doi:10.1038/npp.2015.179, npp2015179 [pii]

Neuromethods (2017) 128: 129–146
DOI 10.1007/7657_2016_9

Published online: 12 August 2016

Experimental Methods for Functional Studies of microRNAs in Animal Models of Psychiatric Disorders

Vladimir Jovasevic and Jelena Radulovic

Abstract

Pharmacological treatments for psychiatric illnesses are often unsuccessful. This is largely due to the poor understanding of the molecular mechanisms underlying these disorders. We are particularly interested in elucidating the mechanism of affective disorders rooted in traumatic experiences. To date, the research of mental disorders in general has focused on the causal role of individual genes and proteins, an approach that is inconsistent with the proposed polygenetic nature of these disorders. We recently took an alternative direction, by establishing the role of miRNAs in the coding of stress-related, fear-provoking memories. Here we describe in detail our work on the role of miR-33 in state-dependent learning, a process implicated in dissociative amnesia, wherein memories formed in a certain brain state can best be retrieved if the brain is in the same state. We present the specific experimental approaches we apply to study the role of miRNAs in this model and demonstrate that miR-33 regulates the susceptibility to state-dependent learning induced by inhibitory neurotransmission.

Keywords: microRNAs, miR-33, Behavior, Learning, Hippocampus, LNA inhibitors, Lentiviral vectors

1 Introduction

Dissociative disorders are thought to arise when normally integrated functions of consciousness, such as memory, perception, or identity awareness, become disrupted by overwhelming stressful experiences. They have a lifetime prevalence rate of about 10 % in the overall population [1]. In addition, dissociative symptoms accompany most psychiatric disorders [2], including PTSD [3] and schizophrenia [4]. These symptoms, which are often debilitating, include partial or complete amnesia, anxiety, depression, suicidality, and severely compromised social functioning [5–7].

Despite the debilitating effects that dissociative disorders have on patients, treatments are frequently unsuccessful because the molecular mechanisms underlying these and all other psychiatric disorders are unknown. To date, research has focused on the causal

The original version of this chapter was revised. An erratum to this chapter can be found at DOI 10.1007/7657_2016_10

role of individual genes and proteins in psychiatric disorders. This approach is complicated by the polygenetic nature of mental disorders, involving not only multiple genes, but also often multiple signaling pathways. Genes often play redundant roles, and therefore, the function of a single gene is masked. Additionally, the main focus so far has been on protein-coding open-reading frames in genomic loci linked to affective disorders, which would miss noncoding RNAs, among which microRNAs are the most prevalent.

MicroRNAs are short RNA molecules, about 22 nucleotides long, that act as posttranscriptional regulators through the inhibition of the target mRNA. They exert their inhibitory activity through the mechanism that involves binding of the microRNA to the complementary sequence usually located within the 3′-untranslated region (3′ UTR) of the target mRNA. Based on the level of complementarity between the microRNA and its target, the target mRNA is either degraded or its translation is blocked [8]. Since the end result of microRNA action is a decrease of protein expression, the term "protein target" is commonly used. Bioinformatics and microarray analyses suggest that the human genome encodes over 1000 microRNAs [9], which are predicted to target about 60 % of human genes [10]. Multiple mRNAs contain the binding site for the same microRNA within their 3′ UTR, which allows one microRNA to silence the expression of many proteins, even when these proteins do not share sequence homology.

MicroRNAs are important for normal physiological function [11–16], as well as in the development and progression of disease [17–24]. The understanding of their role in neuronal processes is beginning to emerge. Many microRNAs localize to specific neuronal compartments, such as the soma, dendrites, or synapses [25–28]. This discrete localization of microRNAs is possibly a part of the cellular mechanism enabling neurons to react to extracellular stimuli at subcellular resolution, even at the level of a single synapse. Several microRNAs are involved in synapse formation and maturation [25, 27, 29], or activity-dependent dendritic remodeling [30, 31]. Learning and memory formation are also controlled by microRNAs, such as miR-132 [32] and miR-134 [33]. Considering their importance in normal brain function, it is likely that dysregulation of brain microRNAs would cause many neuronal diseases, including affective disorders. Several studies have demonstrated a link between altered microRNA expression and psychiatric disorders, identifying microRNAs differentially dysregulated in schizophrenia [34–37], bipolar disorder [35, 36], or major depression [38–40]. Predicted protein targets of these microRNAs are enriched in signaling pathways involved in synaptic function, neurodevelopment, and behavior.

At a fundamental level, dissociative disorders are thought to be rooted in state-dependent encoding of fear-provoking memories, wherein memories encoded in a certain affective or drug-induced state can best be retrieved when the brain is in that same state

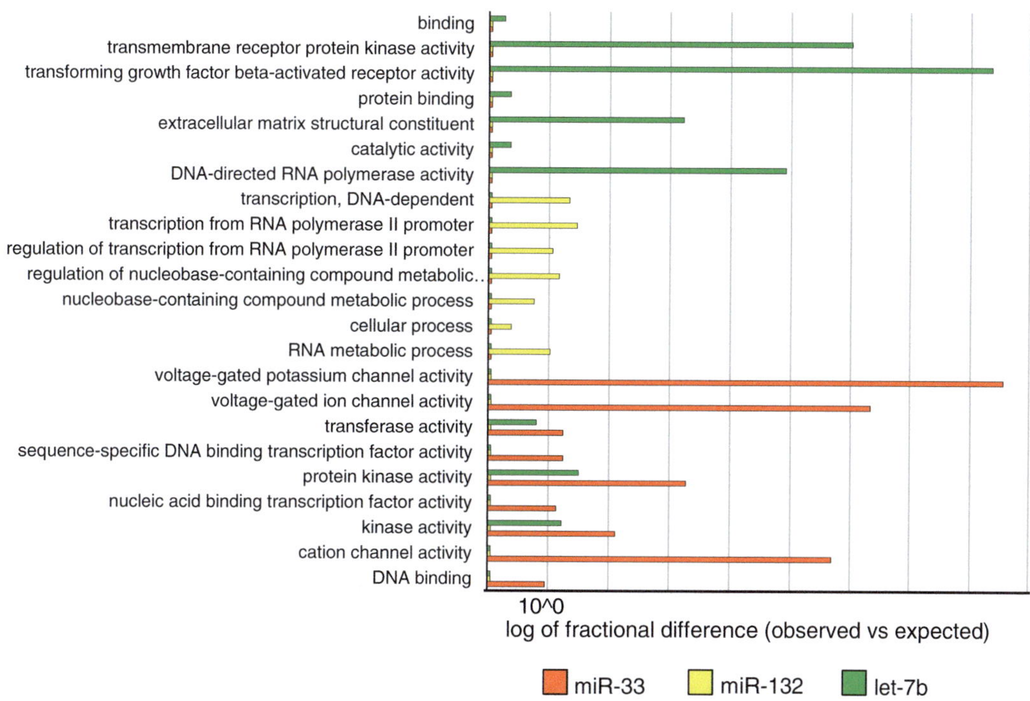

Fig. 9.1 MiRNAs target clusters of functionally related proteins. Predicted targets of miR-33, miR-132 (neuronal) and let-7b (ubiquitous) were analyzed by PANTHER Statistical overrepresentation: GO slim molecular function test and presented as fractional difference of observed vs. expected. Only statistically significant results, $P < 0.05$, are shown

[41, 42]. Whereas significant advances have been made in our understanding of memory processing under normal states of consciousness, state-dependent learning, as it relates to stress-related experiences and their influence on behavior, is not well understood at a neurobiological level. We are interested in deciphering the molecular mechanism of state-dependent processing of fear-provoking memories and understanding how it relates to dissociative disorders. We are particularly interested in understanding how miRNAs regulate state-dependent memories. They fine-tune the amount of their target proteins, and as mentioned, control many functionally related proteins at once (Fig. 9.1). As such, they confer robustness to cellular transitions and states and contribute to gene expression in stable cellular states. We are also interested in miRNAs as a prospective novel therapeutic target.

For the purpose of our studies we developed a mouse model of state-dependent fear, as a putative model of dissociative amnesia. We use contextual fear conditioning as a model for fear-provoking memories, and gaboxadol, an agonist of extrasynaptic $GABA_A$

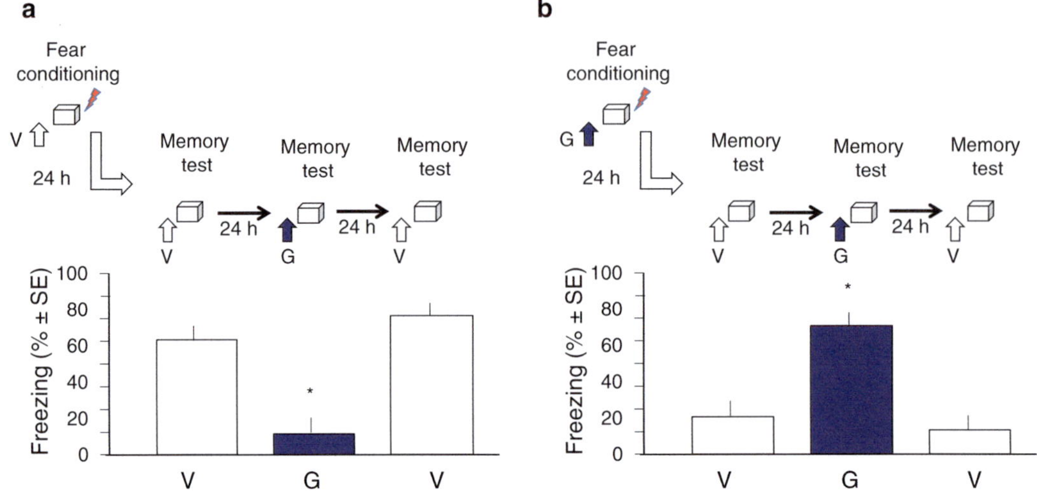

Fig. 9.2 Induction of state-dependent context fear by gaboxadol. (**a**) In mice trained on vehicle (*white*, V), gaboxadol (G, *blue*) impairs memory retrieval. (**b**) In mice trained on gaboxadol, freezing is significantly increased when mice are also tested in the presence of the drug. $^*P < 0.001$ vs. V

receptors, as an inducer of state-dependent learning. When mice are trained on vehicle or gaboxadol and then tested on or off drug on alternate days, in both groups freezing is impaired when mice are tested in a different state (i.e., presence/absence of gaboxadol) from the state in which they were trained, but unaffected if tested in the same state (Fig. 9.2a, b). Using this model we showed that enhanced activity of hippocampal extrasynaptic $GABA_A$ receptors, believed to impair fear and memory, actually enabled their state-dependent encoding and retrieval. This effect requires PKCβII and is influenced by miR-33, a microRNA regulating several GABA-related proteins [43]. An increase or decrease of miR-33 level in the hippocampus did not affect fear conditioning and did not induce the state-dependent learning, but it did change the sensitivity threshold to gaboxadol (Fig. 9.3a). The effect of miR-33 was specific for the regulation of memory and did not affect anxiety-(Fig. 9.3a) and depression-like behavior (Fig. 9.3b).

Here, we describe the unique challenges of the functional studies of miRNAs in intact mouse brain in behavioral paradigms. We describe specific steps we introduced to optimize the experimental protocols. We also share the obstacles we faced in our work and particular approaches we used to overcome them.

2 Materials

2.1 Animal Preparation

Nine-week-old male C57BL/6N mice are obtained from a commercial supplier (Harlan), individually housed on a 12 h light–dark cycle (lights on at 7 a.m.), and allowed ad libitum access to food and water. On the day of arrival, or 1 day later, we surgically implant

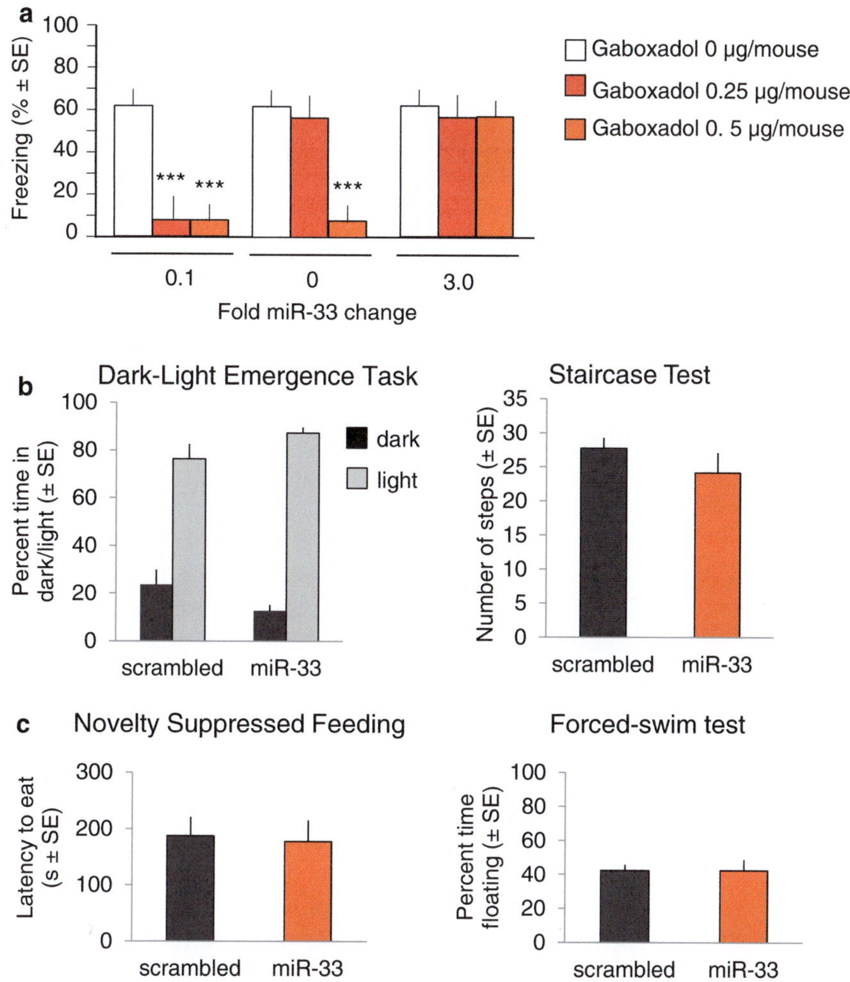

Fig. 9.3 miR-33 determines the susceptibility to gaboxadol-induced state-dependence of fear conditioning. (**a**) Low levels of miR-33 enhance whereas increased levels of this miRNA abolish the effect of gaboxadol on fear conditioning. ***$P < 0.001$ vs. gaboxadol 0 μg/mouse. (**b**) miR-33-LNA does not affect anxiety-like behavior. (**c**) miR-33-LNA does not affect depression-like behavior

cannulas in the selected brain region. We implant double guided cannulas (Plastic One) as described previously [44]. Mice are anesthetized with 1.2 % tribromoethanol (Avertin) and implanted with bilateral 26-gauge cannulas using a stereotaxic apparatus. Since the majority of the studies in our work are done in the hippocampus we provide stereotaxic coordinates for the dorsal hippocampus: 1.7 mm posterior, ±1.0 mm lateral, and 2.0 mm ventral to bregma.

Mice are allowed to rest for 1 week after the surgery prior to the beginning of any pharmacological treatment of behavioral experiment.

2.2 Cells

mHippoE-2 cells (Cedarlane Labs) are maintained in 1× DMEM with 10 % fetal bovine serum (FBS), 25 mM glucose, and 1 % penicillin/streptomycin at 37 °C with 5 % CO_2.

3 Methods

3.1 Detection of miRNAs

We apply a real-time PCR-based protocol for the detection of miRNAs. For the isolation of total RNA in samples from the mouse brain we use miRCURY RNA Isolation Kit-Tissue (Exiqon), with some modification to adjust protocol for the extraction of RNA from the brain tissue. The brain is removed from decapitated animals and immediately placed in ice-cold phosphate-buffered saline (PBS) and the hippocampus isolated from the rest of the brain. Since in most of our studies mice are infused into the dorsal hippocampus, we excise specifically the region of the hippocampus proximal to the site of injection. Usually this encompasses most of the dorsal hippocampus. By doing this we are also able to verify the accuracy of the infusion. Excised tissue is immediately resuspended in the lysis buffer, suspension frozen in liquid nitrogen and stored at −80 °C until RNA extraction. We have also tried the approach in which the tissue itself is flash-frozen in liquid nitrogen and resuspended in the lysis buffer at the time of RNA extraction, and obtained equally high quality RNA and reproducible results. However, frozen tissue is much more difficult to resuspend than the fresh and the process requires much more time, risking possible RNA degradation in the warmed-up tissue. Resuspension of fresh tissue, on the other hand, takes only a few seconds.

We would like also to note that while most of extraction methods would probably work well, it is important to test the efficiency of any method for the specific miRNAs of interest, as some of them fail to extract certain miRNAs. TRIzol extraction protocol, commonly used for total RNA extraction, has been shown inefficient for the extraction of miRNAs with low GC content [45]. We also observed similar phenomenon while using Pure Link RNA Mini Kit (Life Technologies). Mir-33 was selectively lost when using this method, while several other miRNAs were extracted with equal efficiency to Exiqon's kit (Fig. 9.4a). The reason is unlikely to be related to GC content, since miR-33 has similar GC content as some miRNAs that were efficiently extracted using Pure Link RNA Mini Kit (42.85 % for miR-33 vs. 45.45 % for let-7b). The problem of miRNA extraction efficiency is particularly complicated when analyzing miRNAs by high throughput methods. The lack of a miRNA expression can be interpreted as a true result only if the analysis is performed on RNA samples extracted using two different methods.

For the synthesis of cDNA from our RNA samples we use Universal cDNA Synthesis Kit (Exiqon) when analyzing

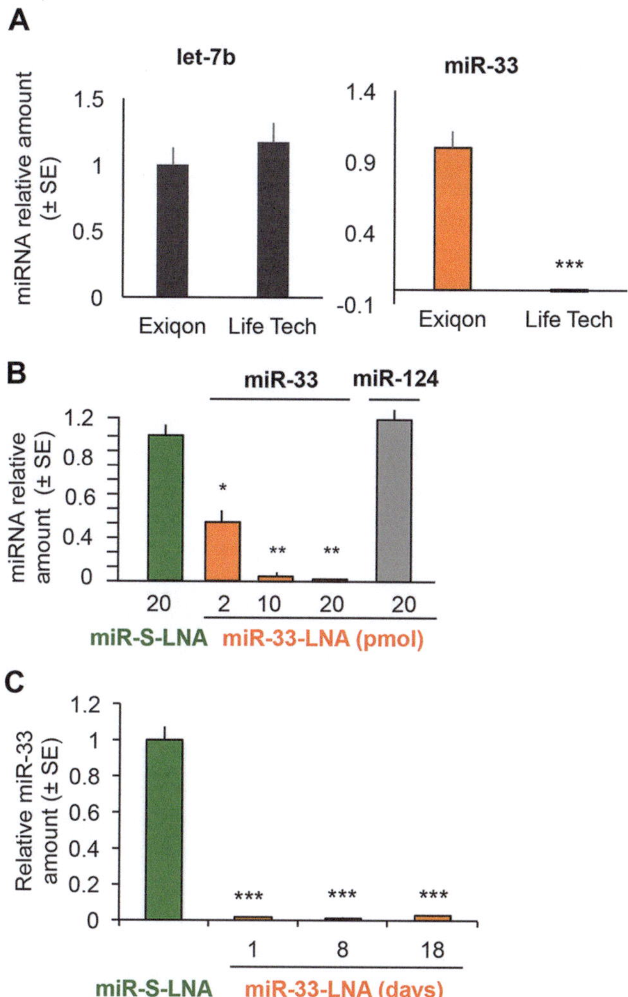

Fig. 9.4 Isolation and downregulation of miR-33. (**a**) Effect on RNA preparation method on miR-33 yield $^{***}P < 0.001$ vs. Exiqon. (**b**) Dose-dependent reduction of miR-33 by intrahippocampal infusion of miR-33-LNA. $^{*}P < 0.05$, $^{**}P < 0.01$ vs. miR-S-LNA (**c**) Lasting effect of a single infusion of miR-33-LNA. $^{***}P < 0.001$ vs. miR-S-LNA

microRNAs, or First Strand cDNA Synthesis Kit (Applied Biosystems) for the analysis of mRNAs. We determined that 20 ng of total RNA gives linear range and a sufficient amount of cDNA for miRNA. When analyzing mRNAs we use 100 ng of total RNA.

We determine the amounts of miRNA by real-time PCR using primers from Exiqon. The primers are locked-nucleic acid (LNA) optimized, which gives very high specificity and extremely low background, allowing the detection of low-abundant miRNAs. We use let-7b miRNA as endogenous control. It is expressed at comparable levels as miR-33, which is mostly the focus of our studies, and its levels do not change in the hippocampus, or other brain regions we examined, in our behavioral experiments. We use standard deviation of all control and experimental samples less than 0.2 as a norm for determining the validity of a miRNA as endogenous control.

Another important parameter is the number of samples for each experimental group. This is particularly the case when studies are done in animals, since the level of miRNAs in naïve state can be very different among animals and can substantially affect the statistical significance of any observed changes. Since the fold change in miRNA levels is often quite low, it can be challenging to detect a significant difference between control and experimental groups. We have determined that in our model four or five animals per each group is usually sufficient to detect significant changes in miRNA levels following our pharmacological treatments. When mice are infused with gaboxadol into the hippocampus we see about 50 % decrease in miR-33 level 1 h after the infusion. Unlike in many other experimental settings, where changes in miRNA levels are transient, the decrease in miR-33 level we observe persist for at least 24 h. Levels of several other miRNAs remain unchanged.

3.2 Inhibition of miRNAs in the Mouse Brain

For the inhibition of miRNAs in selected brain regions, we used locked-nucleic acid (LNA) inhibitors with phosphorothiolated (PS) backbone and 3′ end cholesterol. In these RNA nucleotide analogs the ribose residue is "locked" in the 3′-endo conformation with the addition of a methylene bridge between 2′ oxygen and 4′ carbon, making it assume the ideal conformation for Watson-Crick binding [46]. As a consequence, the melting temperature of duplexes in which one strand is an LNA oligomer increases substantially [47, 48]. This high affinity binding to complementary RNA results in a very high potency when used for antisense inhibition. Another important feature of LNA RNA oligonucleotides is enhanced single nucleotide discrimination and therefore increased target specificity, making them a great tool for targeting of specific miRNAs without affecting other miRNAs, even those with high sequence homology. We were able to determine the amount of inhibitor needed to reduce the level of a miRNA to almost zero, without affecting the level of other miRNAs (Fig. 9.4b). LNA oligonucleotides act rapidly, reaching the full effect as early as 24 h after the intrahippocampal infusion, which was the earliest time point of our studies. It is very likely that the complete inhibition of the target miRNA can be observed sooner than that, considering that a similar approach, utilizing cholesterol-modified oligoribonucleotides complementary to miRNAs (antagomirs), resulted in complete depletion of miR-219 in the suprachiasmatic nucleus 8 h after the intraventricular infusion of miR-219 antagomir [49].

PS modification of the backbone renders the LNA oligonucleotides more resistant to nuclease degradation, effectively enhancing the half-life of the oligonucleotide [50]. Single intrahippocampal infusion of LNA inhibitor completely blocks the target miRNA 18 days after the infusion, which is the latest time point we examined (Fig. 9.4c).

Cholesterol modification at the 3′ end enhances the absorption of the LNA oligonucleotide into the cell, removing the need for transfection reagents.

We specifically use LNA miRNA inhibitors from Exiqon. Inhibitors are dissolved in water to 400 μM concentration and the stock solutions stored at −80 °C. Just prior to each use, the inhibitors are heated at 65 °C for 10 min, cooled on ice, and diluted in artificial cerebrospinal fluid (aCSF) to a final working concentration of 40 μM. The amount of inhibitors we use depends on the site of injection. For the inhibition of miRNAs in the hippocampus we inject 0.5 μl of the working solution into each side (20 pmol of the inhibitor per side). Although for some miRNAs 10 pmol of the inhibitor per side was sufficient for complete inhibition, we decided to use 20 pmol in our experiment since it was effective for all miRNAs we tested so far, and it did not show any off-target effects.

We use as a control a scrambled LNA oligonucleotide, which does not bind to any cellular microRNA. Alternatively, an LNA oligonucleotide with a scrambled target sequence can also be used as a control. However, this approach would require a separate control for each target inhibitor, which in our specific case, where we need at least 8 mice for each experimental group, would unnecessarily add a very large number of animals, making the experiments difficult to manage and very expensive. Although the effect of the LNA inhibitor on target miRNA is rapid, we always wait several days, usually seven, before starting behavioral experiments in order to allow the level of target proteins to be affected by the depletion of miRNA. Many proteins that are the focus of our studies have a relatively long half-life, and in our experience this extended incubation period is required in order to see the increase in the levels of target proteins.

3.3 Overexpression of microRNAs in the Mouse Brain

Two approaches are the most commonly used to increase the levels of a miRNA. One is transfection of miRNA mimic, and the other is transfection of a vector expressing a pre-miRNA-like molecule comprising of miRNA sequence, complementary sequence and a short hairpin loop, or a vector expressing native form of pre-miRNA. Both of the approaches have been modified for in vivo applications and we will discuss them here, with particular emphasis on the vector-based approach.

Transfection of miRNA mimics has advantages for in vitro applications since it results in a rapid increase of cellular miRNA levels. The level remains elevated for a duration of a typical in vitro experiment, usually 48–72 h [51–53]. However, in vivo application of miRNA mimics is more challenging and not often used. In some studies miRNA mimics were mixed with a transfection reagent, as for in vitro use, prior to infusion into the brain [54]. In that study the authors were examining the role of miR-335 in the regulation of cell death in a rat ischemic model and it is possible that the presence of a transfection reagent would not affect the outcome of the experiments. In our system, where we study the role of miRNAs in behavioral paradigms, transfection reagents could likely affect the proper function of neurons and thus the outcome of the experiment. An alternative approach for behavioral studies is infusion of miRNA mimics alone, without a transfection reagent. However, this approach requires multiple infusions before the initiation and through the duration the experiment [55]. We have tested this approach in our studies of the role of miR-33 in learning, and were not able to reliably detect a sustained increase in miR-33 levels in the hippocampus, even after multiple infusions of miR-33 mimic. We did observe a transient increase of miR-33 levels shortly after the infusion into the hippocampus, with a rapid drop to pre-infusion levels (Fig. 9.5a). We do not know if the problem was the poor absorption by hippocampal cells, or the rapid degradation of the mimic. We decided to abandon this approach as unreliable, and instead opted to use expression vector to enhance the levels of miR-33 in the hippocampus.

We use for our studies expression vectors packaged into lentiviral particles. This is a similar system to the one we have used reliably for the overexpression of proteins in different brain regions. The expression vector we use is pMIRNA1/pCDH plasmid encoding miR-33 (System Biosciences). The plasmid consists of the native stem loop structure and 200–400 base pairs of upstream and downstream flanking genomic sequence of miR-33. These

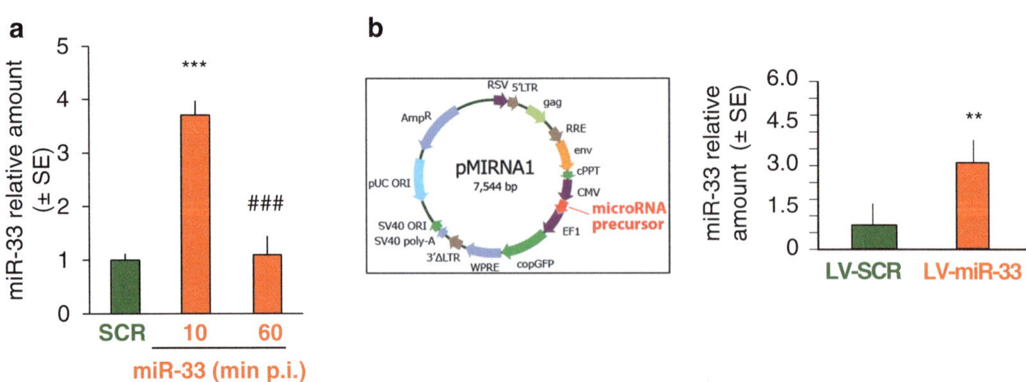

Fig. 9.5 Upregulation of miR-33 levels. (**a**) miR-33 mimic is rapidly lost from the hippocampus. $^{***}P < 0.001$ vs. LV-SCR, $^{\#\#\#}P < 0.001$ vs. miR-33 10 min. (**b**) Lasting upregulation of miR-33 using a lentiviral (LV) vector. Single infusion of miR-33-LNA. $^{**}P < 0.001$ vs. LV-SCR

features ensure that pre-miR-33 expressed from the construct is correctly processed in the cell into mature miR-33. The plasmid also contains copGFP fluorescent marker, optimized for high level expression from the CMV promoter in mammalian cells, which allows the detection of positively transduced cells (Fig. 9.5b, left). The copGFP marker is a green monomeric GFP-like protein. It is nontoxic, non-aggregating protein with fast maturation and high stability. As such, it is very suitable to monitor the spread of the lentiviral particle following the infusion into the brain.

The plasmid is packaged into lentiviral particles at the Northwestern University Genomic Core Facility. For in vivo overexpression, lentiviral particles are injected into the hippocampus at 0.25 μl/side, and allowed to incubate for 30 days before any experiments are performed. Although we can detect the expression from the vector as early as 8 days when we transduce cells in vitro, to get a reliable, substantial expression of miR-33 following infusion into the brain, extended incubation period is needed (Fig. 9.5b, right). The increased amount of miR-33 in the hippocampus we obtain using this approach significantly reduces the level of target proteins.

3.4 Validation of microRNA Targets

Firefly luciferase assay is a common assay to study transcriptional activity. This assay has also been adapted for studies of posttranscriptional regulation of mRNAs by miRNAs, and many different expression systems have been developed for this purpose. When selecting a vector, it is important to keep in mind that a strong promoter, such as CMV promoter, would lead to a high level of expression and would mask the effect of miRNA mimic or miRNA overexpression. Relatively low levels of expression allow changes in both direction to be easily detected. One such vector is pmirGLO Dual-Luciferase miRNA Target Expression Vector (Promega). The expression from the firefly luciferase (*luc2*) reporter gene is driven by human phosphoglycerate kinase (PGK) promoter providing relatively low translational expression. The PGK promoter is a nonviral universal promoter, which functions across cell lines (yeast, rat, mouse, and human). The normalization of gene expression is done relative to humanized Renilla luciferase gene, which is contained within the vector [56].

When it comes to selecting a sequence to be cloned into the vector there are several approaches. Frequently used is cloning the short fragment containing the putative miRNA target sequence from the 3′ UTR [56–59], or even cloning of multiple repeats of the target sequence [60]. This approach allows for a very simplified cloning procedure, often involving synthetic oligonucleotides that can be either digested with appropriate restriction nuclease, or even synthetized with restriction overhangs already present in the generated oligonucleotide. However, this method has several caveats that potentially bring doubts about the validity of the obtained results. First, the cloning of a short segment containing the putative miRNA target sequence from the 3′ UTR creates a very artificial system lacking the rest of the 3′ UTR that can influence the function of a miRNA. Even further, cloning multiple repeats of this sequence might enhance the effect of miRNA on the reporter gene; however, this construct even less resembles the endogenous 3′ UTR of the studied gene. Second, even when the full-length 3′ UTR is cloned, the reporter construct is often transfected into a cell line that does not normally express the gene from which the 3′ UTR comes and thus the effect of a miRNA on the reporter gene is evaluated in a system in which the interaction between the two would not occur naturally. This is particularly the case with neuronally expressed genes, which are the interest of our studies. On a technical note, if luciferase reporter assay is used to verify the targets of a miRNA, with full-length 3′ UTR cloned into the vector, as we believe it should be, longer 3′ UTRs, particularly those with high GC content, could be difficult to amplify by PCR. We faced that problem with several genes we study, such as GABRA4 (1965 bp 3′ UTR), KCC2 (2260 bp), or GABRB2 (5700 bp). For these reasons we opted to identify targets of miRNAs using an assay that relies not on exogenously expressed artificial constructs, but on an interaction of miRNAs with their endogenous target mRNAs.

Biotinylation assay in principle is a simple assay. The miRNA of interest is tagged with biotin, transfected into cells, and tagged miRNA pulled down using streptavidin beads. Target mRNAs are identified either by PCR or a microarray. Biotinylation assay has some advantages over luciferase assay. It relies on binding of miRNAs to their endogenous mRNA targets, in their native form, and it also allows for a high throughput screening and large-scale identification of targets. The assay was originally adapted for miRNA target search by Ørom et al. [61]; however, it suffered from low specificity and poor enrichment of targets [62, 63]. It also involved the introduction of large amounts of miRNA, which could alter the normal function of the cell and alter the stoichiometry of miRNA-mRNA binding, giving false results. In order to keep our studies under as close to natural conditions as possible we decided to use an adapted protocol which resolves the problems of the original method. In this protocol, developed by Wani and Cloonan [64], a small amount of miRNA is transfected into a large number of cells, magnetic beads are blocked with tRNA to reduce nonspecific binding, and several more stringent washes are added to further reduce nonspecific binding. These changes increased the specific enrichment by known and predicted target mRNAs.

An additional difficulty for our studies was the selection of a cell line that would closely resemble the hippocampal neurons, which we study in our mouse model, and which would express the mRNAs of interest. We found mHippoE-2 cells (Cedarlane Labs), an immortalized cell line of mouse primary hippocampal neurons, that express some of the proteins in which we were interested (e.g., KCC2, GABRB4, SYN2a), but not all (GABRA4 was not expressed at reasonable levels). For the identification of miR-33 targets we use biotinylated miRNA mimics (miRIDIAN mimics, Dharmacon) miR-33 (GUGCAUUGUAGUUGCAUUGCA-biotin), and two control miRNAs: scrambled miR-33 in which the miR-33 sequence was scrambled (GGCUCGAUGUUCGAAUAUUGU-biotin), and miR-33mut in which the seed sequence of miR-33 is replaced with the seed sequence of cel-miR-67, a commonly used control miRNA as it does not target known mammalian mRNAs (GGCUCAUUUAGUUGCAUUGCA-biotin). One day before transfection, 1×10^6 cells are seeded onto 10 cm tissue culture dishes (4 dishes for each experimental group). The next day cells are transfected with 50 pmol of biotinylated miRNA per dish using RNAiMAX transfection reagent (Life Technologies). RNAiMAX gives very high transfection efficiency, even in cells known to be very difficult to transfect, such as primary fibroblast [65] or primary neurons [66]. Twenty-four hours later cells are lysed, lysates from the four replicate dishes combined and biotinylated miRNAs pulled down using Dynabeads MyOne Streptavidin C1 (Life Technologies). A portion of the cell lysate is saved as input sample. RNAs obtained in pull-down and input are purified using RNA Easy kit (Quiagen), and concentration measured using NanoDrop (Thermo Scientific). RNA amounts are quantified by real-time PCR. We express the results of the analysis as the enrichment ratio of the control-normalized pull-down RNA to the control-normalized input levels, and present them relative to miR-33 group, as originally done by Ørom et al. [61]. Using this modified assay we were able to verify KCC2 and GABRB2, predicted to be targets of miR-33 by bioinformatics analysis, and whose protein levels change with manipulations of miR-33 levels (Fig. 9.6a), as true targets of miR-33 (Fig. 9.6b). Syn2a, whose protein levels also changed in response to miR-33 inhibition and overexpression, is an indirect target, since its mRNA is not pulled down with miR-33.

4 Conclusions

MiRNAs are emerging as important regulators of physiological functions under normal conditions, and their dysregulation is increasingly associated with major human diseases [17–24]. Their levels are commonly found perturbed in patients suffering from a range of psychiatric illnesses [34–40], including miR-33, consistently dysregulated in patients suffering from major depression and psychosis [36, 67]. Inhibitors of miRNAs and vector-based overexpression systems provided the means of functional studies in cell culture systems. Novel generations of inhibitors and overexpression systems enabled the translation of cell culture studies into in vivo

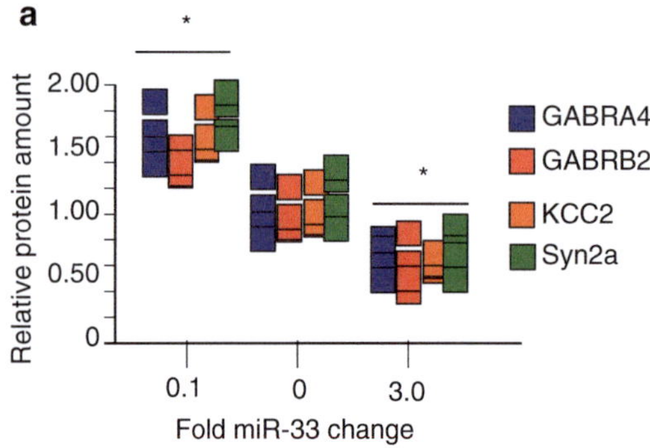

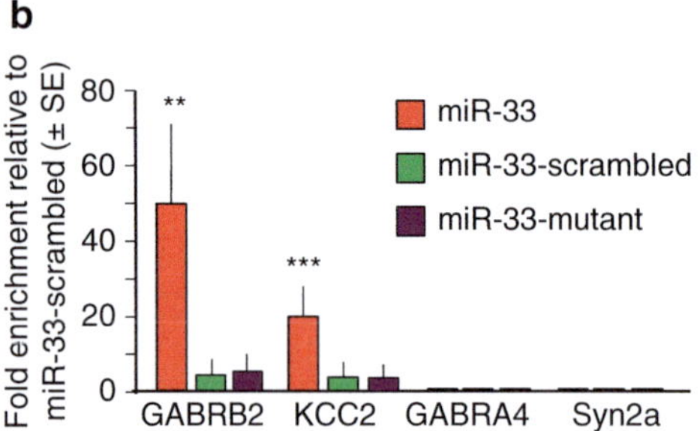

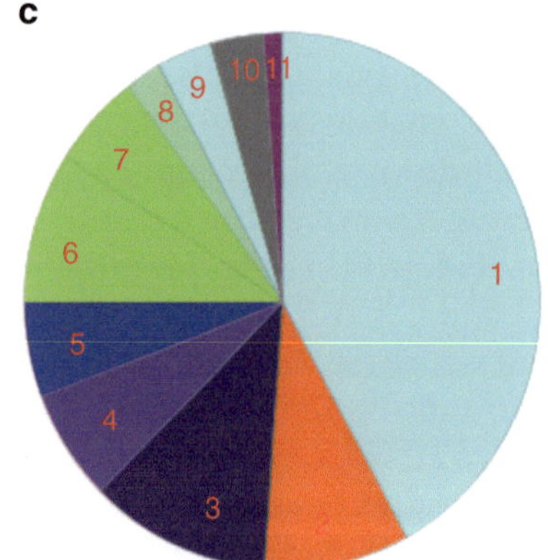

1 G-protein coupled receptor activity
2 GABA receptor activity
3 Acetylcholine receptor activity
4 Cytokine receptor activity
5 Glutamate receptor activity
6 Ligand-activated sequence-specific DNA binding RNA polymerase II transcription factor activity
7 Transmembrane receptor protein serine/threonine kinase activity
8 Transmembrane receptor protein serine/threonine kinase activity
9 Transmembrane receptor protein tyrosinekinase activity
10 Tumor necrosis factor receptor binding
11 Tumor necrosis factor activated receptor activity

Fig. 9.6 Validation of miR-33 targets. (**a**) Changes of miR-33 levels inversely affect the levels of GABA-related proteins targeted by miR-33. $^{*}P < 0.05$, vs. control (0). (**b**) Cell line mHippoE-2 expressing GABRB2 and KCC2 (but not GABRA4) show that these mRNAs are co-eluted with biotynilated miR-33. $^{**}P < 0.01$, $^{***}P < 0.001$ vs. control miRNAs. (**c**) GABA-related proteins are among the prominent targets of miR-33 (PANTHER analysis of molecular function, cluster 6: receptor function)

animal models, without the need for transfection reagents, keeping the experimental conditions as close to normal physiological conditions as possible. By combining behavioral paradigms with miRNA manipulations in animal brain we are now able to address the role miRNAs have in normal brain function and the development of mental disorders. We believe that the studies of miRNAs in the context of psychiatric disorders will also lead to novel, miRNA-based therapies for these disorders.

The development of drugs for the treatment of psychiatric disorders has often been unsuccessful. Drugs that have a broad spectrum have serious side effects because they target many proteins irrelevant for these disorders, but important for normal cellular function. On the other hand, drugs that are highly selective for one or very few proteins are often ineffective due to the fact that affective disorders are complex and involve multiple proteins and mechanisms. Focusing therapeutic approaches onto miRNAs instead of proteins offers several benefits:

1. By therapeutically targeting single miRNA it is possible to modify the level of many proteins simultaneously. These proteins will likely be clustered within specific pathways relevant for the development and progression of the disease, providing at the same time broad and selective effects.
2. The effect miRNAs have on the protein level of their targets is very modest [68, 69]. As the result, miRNA-based therapies would have minimal side effects since there would be no dramatic changes of a single protein, but would have a strong overall effect due to the above mentioned simultaneous regulation of numerous targets.
3. Unlike proteins, miRNAs are very easily targeted, with very high specificity, either by the use of highly specific inhibitors or artificial miRNA mimics, depending on whether the goal is the decrease or increase in microRNA level.

We expect that with these benefits in mind, and with feasibility of miRNA functional research today, we will see a major shift towards miRNA-based or miRNA-targeting therapies of mental illnesses.

Acknowledgments

This work was supported by NIH grants NIMH/MH078064 (J.R.) and Ken and Ruth Davee Award for Innovative Investigations in Mood Disorders, (J.R. and V.J.).

References

1. Sar V (2011) Developmental trauma, complex PTSD, and the current proposal of DSM-5. Eur J Psychotraumatol 2:5622. doi: 10.3402/ejpt.v2i0.5622
2. Sar V, Ross C (2006) Dissociative disorders as a confounding factor in psychiatric research. Psychiatr Clin North Am 29(1):129–144, ix
3. Lanius RA et al (2010) Emotion modulation in PTSD: clinical and neurobiological evidence for a dissociative subtype. Am J Psychiatry 167(6):640–647
4. Ross CA, Keyes BB (2004) Dissociation and schizophrenia. J Trauma Dissociation 5:69–83
5. Dorahy MJ et al (2013) Complex trauma and intimate relationships: the impact of shame, guilt and dissociation. J Affect Disord 147 (1–3):72–79
6. Krammer S, Kleim B, Simmen-Janevska K, Maercker A (2015) Childhood trauma and complex posttraumatic stress disorder symptoms in older adults: a study of direct effects

and social-interpersonal factors as potential mediators. J Trauma Dissociation 26:1–16

7. Renard SB, Pijnenborg M, Lysaker PH (2012) Dissociation and social cognition in schizophrenia spectrum disorder. Schizophr Res 137 (1–3):219–223
8. Bartel DP (2004) MicroRNAs: genomics, biogenesis, mechanism, and function. Cell 116 (2):281–297
9. Bentwich I et al (2005) Identification of hundreds of conserved and nonconserved human microRNAs. Nat Genet 37 (7):766–770
10. Friedman RC, Farh KK, Burge CB, Bartel DP (2009) Most mammalian mRNAs are conserved targets of microRNAs. Genome Res 19 (1):92–105
11. Ambros V (2004) The functions of animal microRNAs. Nature 431(7006):350–355
12. Chen X (2004) A microRNA as a translational repressor of APETALA2 in Arabidopsis flower development. Science 303(5666):2022–2025
13. Lee RC, Feinbaum RL, Ambros V (1993) The C. elegans heterochronic gene lin-4 encodes small RNAs with antisense complementarity to lin-14. Cell 75(5):843–854
14. Qi J et al (2009) microRNAs regulate human embryonic stem cell division. Cell Cycle 8 (22):3729–3741
15. Reinhart BJ et al (2000) The 21-nucleotide let-7 RNA regulates developmental timing in Caenorhabditis elegans. Nature 403(6772): 901–906
16. Zhang X, Wan G, Berger FG, He X, Lu X (2011) The ATM kinase induces microRNA biogenesis in the DNA damage response. Mol Cell 41(4):371–383
17. Li R et al (2010) MicroRNAs involved in neoplastic transformation of liver cancer stem cells. J Exp Clin Cancer Res 29:169
18. Cittelly DM et al (2010) Downregulation of miR-342 is associated with tamoxifen resistant breast tumors. Mol Cancer 9:317
19. Rederstorff M, Huttenhofer A (2010) Small non-coding RNAs in disease development and host-pathogen interactions. Curr Opin Mol Ther 12(6):684–694
20. Mestdagh P et al (2010) The miR-17-92 microRNA cluster regulates multiple components of the TGF-beta pathway in neuroblastoma. Mol Cell 40(5):762–773
21. Moriyama T et al (2009) MicroRNA-21 modulates biological functions of pancreatic cancer cells including their proliferation, invasion, and chemoresistance. Mol Cancer Ther 8 (5):1067–1074
22. Callis TE et al (2009) MicroRNA-208a is a regulator of cardiac hypertrophy and conduction in mice. J Clin Invest 119(9):2772–2786
23. Dong S et al (2009) MicroRNA expression signature and the role of microRNA-21 in the early phase of acute myocardial infarction. J Biol Chem 284(43):29514–29525
24. Fulci V et al (2010) miR-223 is overexpressed in T-lymphocytes of patients affected by rheumatoid arthritis. Hum Immunol 71 (2):206–211
25. Schratt GM et al (2006) A brain-specific microRNA regulates dendritic spine development. Nature 439(7074):283–289
26. Kye MJ et al (2007) Somatodendritic microRNAs identified by laser capture and multiplex RT-PCR. RNA 13(8):1224–1234
27. Siegel G et al (2009) A functional screen implicates microRNA-138-dependent regulation of the depalmitoylation enzyme APT1 in dendritic spine morphogenesis. Nat Cell Biol 11 (6):705–716
28. Lugli G, Torvik VI, Larson J, Smalheiser NR (2008) Expression of microRNAs and their precursors in synaptic fractions of adult mouse forebrain. J Neurochem 106(2):650–661
29. Caygill EE, Johnston LA (2008) Temporal regulation of metamorphic processes in Drosophila by the let-7 and miR-125 heterochronic microRNAs. Curr Biol 18(13):943–950
30. Wayman GA et al (2008) An activity-regulated microRNA controls dendritic plasticity by down-regulating p250GAP. Proc Natl Acad Sci U S A 105(26):9093–9098
31. Vessey JP et al (2006) Dendritic localization of the translational repressor Pumilio 2 and its contribution to dendritic stress granules. J Neurosci 26(24):6496–6508
32. Hansen KF, Sakamoto K, Wayman GA, Impey S, Obrietan K (2010) Transgenic miR132 alters neuronal spine density and impairs novel object recognition memory. PLoS One 5(11): e15497
33. Gao J et al (2010) A novel pathway regulates memory and plasticity via SIRT1 and miR-134. Nature 466(7310):1105–1109
34. Perkins DO et al (2007) microRNA expression in the prefrontal cortex of individuals with schizophrenia and schizoaffective disorder. Genome Biol 8(2):R27
35. Kim AH et al (2010) MicroRNA expression profiling in the prefrontal cortex of individuals affected with schizophrenia and bipolar disorders. Schizophr Res 124(1–3):183–191
36. Moreau MP, Bruse SE, David-Rus R, Buyske S, Brzustowicz LM (2011) Altered microRNA expression profiles in postmortem brain

samples from individuals with schizophrenia and bipolar disorder. Biol Psychiatry 69 (2):188–193

37. Cammaerts S et al (2015) Schizophrenia-associated MIR204 regulates noncoding RNAs and affects neurotransmitter and ion channel gene sets. PLoS One 10(12): e0144428
38. Saus E et al (2010) Genetic variants and abnormal processing of pre-miR-182, a circadian clock modulator, in major depression patients with late insomnia. Hum Mol Genet 19 (20):4017–4025
39. Xu Y et al (2010) A polymorphism in the microRNA-30e precursor associated with major depressive disorder risk and P300 waveform. J Affect Disord 127(1–3):332–336
40. Liang Y et al (2015) Genetic variants in the promoters of let-7 family are associated with an increased risk of major depressive disorder. J Affect Disord 183:295–299
41. Braun BG (1984) Towards a theory of multiple personality and other dissociative phenomena. Psychiatr Clin North Am 7(1):171–193
42. Spiegel D et al (2011) Dissociative disorders in DSM-5. Depress Anxiety 28(9):824–852
43. Jovasevic V et al (2015) GABAergic mechanisms regulated by miR-33 encode state-dependent fear. Nat Neurosci 18(9): 1265–1271
44. Radulovic J, Sydow S, Spiess J (1998) Characterization of native corticotropin-releasing factor receptor type 1 (CRFR1) in the rat and mouse central nervous system. J Neurosci Res 54(4):507–521
45. Kim YK, Yeo J, Ha M, Kim B, Kim VN (2011) Cell adhesion-dependent control of microRNA decay. Mol Cell 43(6):1005–1014
46. Obika S et al (1997) Synthesis of 2′-O,4′-C-methyleneuridine and -cytidine. Novel bicyclic nucleosides having a fixed C-3,-endo sugar puckering. Tetrahedron Lett 38 (50):8735–8738
47. Kaur H, Arora A, Wengel J, Maiti S (2006) Thermodynamic, counterion, and hydration effects for the incorporation of locked nucleic acid nucleotides into DNA duplexes. Biochemistry 45(23):7347–7355
48. Owczarzy R, You Y, Groth CL, Tataurov AV (2011) Stability and mismatch discrimination of locked nucleic acid-DNA duplexes. Biochemistry 50(43):9352–9367
49. Cheng HY et al (2007) microRNA modulation of circadian-clock period and entrainment. Neuron 54(5):813–829
50. Vosberg HP, Eckstein F (1982) Effect of deoxynucleoside phosphorothioates incorporated in DNA on cleavage by restriction enzymes. J Biol Chem 257(11):6595–6599
51. Othumpangat S, Walton C, Piedimonte G (2012) MicroRNA-221 modulates RSV replication in human bronchial epithelium by targeting NGF expression. PLoS One 7(1): e30030
52. Pal R, Greene S (2015) microRNA-10b is overexpressed and critical for cell survival and proliferation in medulloblastoma. PLoS One 10 (9):e0137845
53. Wang S et al (2015) The potent tumor suppressor miR-497 inhibits cancer phenotypes in nasopharyngeal carcinoma by targeting ANLN and HSPA4L. Oncotarget 6(34): 35893–35907
54. Liu FJ et al (2015) MiR-335 regulates Hif-1alpha to reduce cell death in both mouse cell line and rat ischemic models. PLoS One 10(6): e0128432
55. Zovoilis A et al (2011) microRNA-34c is a novel target to treat dementias. EMBO J 30 (20):4299–4308
56. Guo R, Abdelmohsen K, Morin PJ, Gorospe M (2013) Novel microRNA reporter uncovers repression of let-7 by GSK-3beta. PLoS One 8(6):e66330
57. Jiang L et al (2010) Downregulation of the Rho GTPase signaling pathway is involved in the microRNA-138-mediated inhibition of cell migration and invasion in tongue squamous cell carcinoma. Int J Cancer 127 (3):505–512
58. Wu T et al (2012) miR-30 family members negatively regulate osteoblast differentiation. J Biol Chem 287(10):7503–7511
59. Krishnan K et al (2013) MicroRNA-182-5p targets a network of genes involved in DNA repair. RNA 19(2):230–242
60. Shi L, Ko ML, Ko GY (2009) Rhythmic expression of microRNA-26a regulates the L-type voltage-gated calcium channel alpha1C subunit in chicken cone photoreceptors. J Biol Chem 284(38):25791–25803
61. Orom UA, Lund AH (2007) Isolation of microRNA targets using biotinylated synthetic microRNAs. Methods 43(2):162–165
62. Nonne N, Ameyar-Zazoua M, Souidi M, Harel-Bellan A (2010) Tandem affinity purification of miRNA target mRNAs (TAP-Tar). Nucleic Acids Res 38(4):e20
63. Orom UA, Nielsen FC, Lund AH (2008) MicroRNA-10a binds the 5′UTR of ribosomal protein mRNAs and enhances their translation. Mol Cell 30(4):460–471
64. Wani S, Cloonan N (2014) Profiling direct mRNA-microRNA interactions using synthetic

biotinylated microRNA-duplexes. bioRxiv. doi: 10.1101/005439

65. Jovasevic V, Naghavi MH, Walsh D (2015) Microtubule plus end-associated CLIP-170 initiates HSV-1 retrograde transport in primary human cells. J Cell Biol 211(2):323–337
66. Han Z et al (2015) Establishment of lipofection protocol for efficient miR-21 transfection into cortical neurons in vitro. DNA Cell Biol 34(12):703–709
67. Banigan MG et al (2013) Differential expression of exosomal microRNAs in prefrontal cortices of schizophrenia and bipolar disorder patients. PLoS One 8(1):e48814
68. Baek D et al (2008) The impact of microRNAs on protein output. Nature 455(7209):64–71
69. Hendrickson DG et al (2009) Concordant regulation of translation and mRNA abundance for hundreds of targets of a human microRNA. PLoS Biol 7(11):e1000238

Neuromethods (2017) 128: 147–159
DOI 10.1007/7657_2016_8

Published online: 12 August 2016

Isolation and Quantitative Analysis of Axonal Small Noncoding RNAs

Hak Hee Kim, Monichan Phay, and Soonmoon Yoo

Abstract

The success of axonal regeneration of injured neurons depends largely on the intrinsic capacity for intra-axonal protein synthesis that is precisely regulated in a spatial and temporal manner. Recent studies have uncovered important roles of small noncoding RNAs in injury responses of axons during regeneration, particularly via translational regulation of axonally localized mRNAs. Several different approaches have been primarily focused on in vitro such as a modified Boyden chamber, a Campenot chamber, and a microfluidic device to study the miRNA profiles of axonal compartments because of contamination from non-neuronal cells in vivo. However, the in vivo studies include the ability to reflect all the possible consequences of reciprocal cellular interactions of the reinnervating injury axons and the denervated targets that cannot be replicated faithfully in culture. Here, we discuss an in-depth method for isolating axoplasm from rat sciatic nerve to purity for analyses of axonal miRNA content. We further show how this method can be used to quantitatively analyze specific miRNAs whose levels in axon is altered in response to injury, providing a means to understand how intra-axonal protein synthesis is regulated during regenerative processes. Reliable and reproducible methodologies to purify axoplasm from whole nerve would ultimately provide novel mechanistic insights into axonal injury and regeneration.

Keywords: RNA transport, Intra-axonal protein synthesis, Small noncoding RNAs, Axonal miRNA

1 Introduction

Disturbance of structural and functional connectivity in the nervous system is a devastating consequence of traumatic nerve injury as well as many neurodegenerative diseases. Because of the inherent low capacity of central neurons to regenerate, axon loss in the central nervous system (CNS) usually leads to permanent neurophysiological dysfunction [1–3]. By contrast, spontaneous axon regeneration occurs after injury to the peripheral nervous system (PNS) [4, 5]. Although the exact cellular/molecular mechanisms that induce the regenerative responses in the PNS still remain to be elucidated, growing evidence indicates that successful axon regeneration of adult PNS neurons relies on a sophisticated coordination of gene expression in both the cell body and axon proximal to the injury [6–8]. Determining molecular mechanisms that lead to regeneration in the PNS will identify new pathways that can be therapeutically targeted for CNS injury and potentially other degenerative diseases in humans.

Peripheral nerve injuries trigger rapid and local regenerative responses for which local translation of mRNAs in axons is essential to initiate damage repair, to form a new growth cone and to generate retrograde transporting injury signal [9–19]. This intra-axonal protein synthesis must be regulated precisely in a spatial and temporal manner. Spatial regulation is achieved in large part by regulating the transport of mRNAs into axons [20–23]. However, it is also clear that temporal regulation occurs through translational regulation of localized mRNAs. For example, PNS nerve injury triggers a rapid translation of *importin-β1*, *vimentin*, *RanBP1*, and *Stat3α* mRNAs in axons [9–13]. The proteins encoded by these axonal mRNAs provide retrograde injury signals to the cell body. Other axonally synthesized proteins contribute to growth cone formation [14]. All available data indicate that these mRNAs are delivered to axons from the cell body and that intra-axonal translation of these mRNAs is precisely regulated as needed for autonomous response to local stimuli including nerve injury. However, it is not currently known how this intra-axonal translation is initiated, nor how this activated translation is downregulated once nerve repair is complete. Small non-coding RNA (sncRNA) pathways are strong candidates as controllers of this translational switch.

Over the past decade, microRNAs (miRNAs) have emerged as key regulators in the control of gene expression in the nervous system [24–27]. The primary action of miRNAs is to negatively regulate gene expression by binding to target mRNAs, typically binding to the 3′ untranslated region (UTR) of the target mRNAs [28, 29]. For example, in a spinal muscular atrophy (SMA) mouse model in which neurons are deficient in the survival of motor neuron protein (SMN), axonal levels of miR-183 are greatly increased compared with control mice [30]. The increased level of miR-183 downregulates the axonal protein synthesis of *mTor* (the mammalian target of rapamycin) by direct binding to the 3′UTR of the mRNA. The consequent reduction in local axonal translation of *mTor* could cause an overall reduction in local protein synthesis of SMN in growing neurons, leading to neuronal disorders and aberrant neuronal development. The suppression of miR-183 using miRNA sponges carrying its potential target sequence in vivo improves some neurophysiological defects in an SMA mouse model [30]. Similar studies have shown that the miR-17–92 cluster downregulates synthesis of PTEN protein in axons of embryonic cortical neurons and that miR-16 downregulates synthesis of the eukaryotic translation initiation factors, eIF2B2 and eIF4G2 proteins, in sympathetic axons [31–33]. The elucidation of control of local protein synthesis within the axonal compartment at different levels of neuronal activity could potentially lead us to develop RNA-based approaches for the therapy of neurological disorders.

Recent advances in the analysis of gene expression at unprecedented sensitivity, depth and accuracy greatly accelerate our understanding of the molecular mechanisms involved in regulation of gene expression and networks for axon regeneration [34–36]. Concomitant with the advanced development of high-throughput technologies including next-generation sequencing and increased use of computational tools for transcriptome profiling analysis, an increasing number of studies have investigated the differential distribution and temporal changes of miRNA levels in distal processes of neurons during regeneration. In vitro, physical separation of distal processes of PNS and CNS neurons from their cell bodies have facilitated our understanding of the spatiotemporal dynamics of miRNA levels in regenerating processes of neurons [33, 37–41]. Although this in vitro system has been successfully used to profile differential levels of axonal miRNAs between naïve and regenerating neurons, the in vivo approach further includes the ability to reflect all of the possible consequences of reciprocal cellular interactions of the reinnervating injured axons and the denervated targets that cannot be replicated faithfully in culture [40, 42, 43]. However, previous in vivo studies of high-throughput transcriptome analysis of distal processes of neurons could be confounded by potential contamination from non-neuronal cells during microdissection of nerve tissues and axonal RNA isolation [39, 43]. Therefore, a more specific isolation of axonal RNA content from in vivo preparation, ideally pure intra-axonal RNAs, is needed to study critical molecular regulators of spatiotemporal-specific gene expression that is directly linked to neurodevelopment, neurodegenerative disorders, and regeneration. Here we introduce the methodology that we have optimized both for profiling and quantification of axonal small noncoding RNA contents that are differentially responding to injury in regenerating axons.

2 Materials

1. Sprague Dawley Rats (male, 150 g).
2. RNase decontamination solution (Ambion #AM9780).
3. Nuclease-free phosphate buffered saline without calcium and magnesium, pH 7.4 (Ambion #AM9624).
4. 60 mm cell culture dish (Falcon #353802).
5. Dumont #5 fine forceps (Fine Science Tools #11254-20).
6. Stereo microscope (Leica S6 D with L2).
7. Surgical blade #20 (Bard-Parker #371220).
8. Nuclease-free microcentrifuge tubes (USA scientific 1.5 mL #1615-5510, 1.5 mL Amber # 1615-5507, 2.0 mL #1620-2700).

9. TRIzol reagent (Life Technologies #15596026).
10. RNase-free disposable pellet pestle (Fisher #12-141-364).
11. Chloroform (molecular biology grade, MP Biomedicals #0219400280).
12. Isopropanol (Fisher Bioreagents #BP2618500).
13. Ethanol (Fisher Bioreagents #BP2818500).
14. Ultrapure Glycogen (Invitrogen #10814-010).
15. miRNeasy mini kit (QIAGEN #217004).
16. RNeasy MinElute Cleanup kit (QIAGEN #74204).
17. Nuclease-free water (IDT #11-05-01-14).
18. Quant-iT RiboGreen RNA Assay Kit (Invitrogen #R11490).
19. Nuclease-free 1× TE buffer pH 8.0 (IDT #11-05-01-13).
20. VersaFluor Microcuvette (Bio-Rad #1702416).
21. VersaFluor Fluorometer System (Bio-Rad #170-2402) with filter Set (EX 490/10, EM 520/10) (BioRad #1702436).
22. iScript cDNA Synthesis Kit (Bio-Rad #170-8891).
23. NCode VILO miRNA cDNA Synthesis Kit (Life Technologies #A11193-050).
24. HotStarTaq Master Mix Kit (Qiagen #203443).
25. NCode EXPRESS SYBR GreenER miRNA qRT-PCR Kit (Life Technologies #A11193-051).
26. Real-time primers for each transcript of interest optimized for comparative cycle threshold (C_t) method (*see* **Notes 17** and **18**).
27. Biomek 3000® Laboratory Automation Workstation (Beckman Coulter).
28. MicroAmp® Optical 384-Well Reaction Plate with Barcode (Life Technologies, #4309849).
29. MicroAmp® Optical Adhesive Film (Life Technologies, #4311971).
30. ABI 7900HT Real-Time PCR System (Applied Biosystems).

3 Methods

3.1 Mechanical Squeezing Method for Axoplasm Extraction from Sciatic Nerve

3.1.1 Sciatic Nerve Dissection and Process

1. Dissect out sciatic nerves (~3 cm in length) from rats and place the nerves into prechilled 2.0 mL tube on ice (*see* **Note 1**).
2. Wash the nerves twice with ice-cold nuclease-free PBS and transfer the nerves to a 60 mm petri dish containing 2 mL of ice-cold nuclease free PBS on ice. The following procedures should be done on ice.

3. Remove non-axonal connective tissues from the dissected nerves using two pairs of fine forceps (*see* **Note 2**).
4. Cut and remove 0.5 cm of both ends of the nerve using surgical blade and transfer the remaining nerve to a new petri dish with 2 mL PBS.
5. Cut the nerve into 5 mm segments using surgical blade.

3.1.2 Axoplasm Extraction by Mechanical Squeezing Method

1. Place the nerve segment into a 1.5 mL tube containing 0.2 mL of TRIzol reagent (*see* **Note 3**).
2. Push the nerve down to the TRIzol reagent with a pestle slowly so that the sciatic nerve segment does not float.
3. Apply a quick but gentle force to the sciatic nerve segment to squeeze out axoplasm into the TRIzol reagent 2–3 times (*see* **Note 4**).
4. Remove the processed nerve segment out of the tube immediately and add a new tissue segment.
5. Repeat the mechanical squeezing for the rest of the sciatic nerve segments (*see* **Note 5**).
6. Adjust the final volume of TRIzol to 1 mL (*see* **Note 6**).

3.1.3 RNA Isolation

1. Incubate the sciatic nerve axoplasm in TRIzol reagent at room temperature for 5 min.
2. Add 0.2 mL of chloroform and shake the tube vigorously by hand for 15 s.
3. Incubate the tube for 3 min at room temperature.
4. Centrifuge the tube at 12,000 × *g* for 15 min at 4 °C.
5. Transfer the upper aqueous phase (approximately 600 μL) to a new 1.5 mL tube.
6. For Total RNA extraction, proceed to step 7 below. For small RNA fractionation, proceed to Subheading 3.1.4 (*see* **Note** 7).
7. Add 0.5 mL of 100 % isopropyl alcohol (per 1 mL of TRIzol) and 10 μg of nuclease-free glycogen followed by incubating for 10 min at room temperature (*see* **Note 8**).
8. Centrifuge at 12,000 × *g* for 10 min at 4 °C.
9. Remove supernatant and wash the pellet with 75 % ethanol.
10. Centrifuge again and remove supernatant. Air-dry the pellet on ice for 5 min.
11. Suspend RNA with nuclease-free water. DNase can be treated prior to reverse transcription.

3.1.4 Small RNA Fractionation

1. Please review manufacturer's manual carefully before starting small RNA fractionation even though general procedures are

described below. RNAs smaller than 200 nucleotides will be enriched into fraction.

2. Add 1 volume (600 μL) of 70 % nuclease-free ethanol to the aqueous phase obtained above and load the sample into an RNeasy Mini spin column placed in a 2.0 mL collection tube followed by centrifuging at ≥8000 × *g* for 15 s at room temperature (*see* **Note 9**).
3. Collect and transfer the flow-through to a new nuclease-free 2 mL tube, and save the RNeasy Mini spin column on ice (*see* **Note 10**).
4. Add 0.78 mL of 100 % ethanol (0.65 volume of the flow-through) and mix thoroughly with a brief vortex.
5. Load the sample onto an RNeasy MinElute spin column placed in a 2 mL collection tube.
6. Centrifuge at ≥8000 × *g* for 15 s at room temperature and discard the flow-through (*see* **Note 10**).
7. Add 0.5 mL Buffer RPE in the RNeasy MinElute Kit into the spin column and centrifuge at ≥8000 × *g* for 15 s at room temperature and discard the flow-through.
8. Load 0.5 mL of 80 % ethanol to the spin column and centrifuge at ≥8000 × *g* for 2 min at room temperature.
9. Transfer the RNeasy MinElute spin column into a new 2.0 mL tube and centrifuge at ≥8000 × *g* for 5 min at room temperature with the lid opened.
10. Transfer the spin column into a new nuclease-free 1.5 mL tube and elute the small RNA fraction into nuclease-free water (20 μL total elution volume) by centrifuging at ≥8000 × *g* for 5 min at room temperature.

3.1.5 RNA Quantification and Reverse Transcription

1. Quantify 2 μL of each RNA sample using Quanti-iT Ribo-Green Assay Kit with high-range working solution in a total volume of 200 μL (*see* **Note 11**).
2. For generating a standard curve, dilute the original ribosomal RNA standard (100 μg/mL) from the Kit 1000-fold in 1× TE to make the 100 ng/mL ribosomal RNA.
3. Add 100 μL of RNA standards in five different concentrations in 1× TE to each amber microcentrifuge tube (*see* **Note 12**).
4. Add 100 μL of diluted RNA samples in 1× TE to each amber microcentrifuge tubes.
5. Add 100 μL of the RiboGreen working solution into the tubes containing standards or samples. Vortex briefly and transfer the solution into each VersaFluor cuvettes.

6. Read RNA-RiboGreen fluorescence on a VersaFluor fluorometer with the excitation 485–495 nm and emission 515–525 nm filters per quantification kit manufacturer's protocol (*see* **Note 13**).
7. Calculate the RNA concentration using standard curve.
8. Use approximately 50 ng of total RNA for the reverse transcription (RT) reaction using the iScript cDNA synthesis kit (*see* **Note 14** for axonal purity check). Use NcodeVILO cDNA synthesis kit for small RNAs to incorporate poly (A) tailing (*see* **Note 15**).

3.1.6 Quantitative Real-Time PCR (qRT-PCR)

The method outlined below is based on relative quantification for real-time analysis with SYBR Green, and the reaction condition is optimized for Biomek 3000® Laboratory Automation Workstation and Applied Biosystems 7900HT Fast Real-Time PCR System. This reaction condition should be optimized depending on the system utilized for qRT-PCR.

1. Set up the qRT-PCR reaction using the NCode™ EXPRESS SYBR® GreenER™ qRT-PCR SuperMix Universal for sncRNAs.
2. A robotic system (Biomek 3000) is used to standardize pipetting of the qPCR reaction into the 384-well plate.
3. Apply a MicroAmp Optical Adhesive Film to plate and spin down at 100 × *g* for 1 min.
4. Define the plate layout and reaction cycling condition in SDS 2.3 (*see* **Note 16**) and initiate the qPCR running.
5. When the run has finished, recover the plate and proceed with data analysis. The automatic comparative threshold (C_t) algorithm of the ABI Prism SDS 2.3 and RQ manager software (Applied Biosystem Software) is used to calculate the optimal baseline range and threshold values for an endogenous control and each transcript of interests (*see* **Note 17**).

3.2 Direct Contrast of Axonal Levels of miRNAs Before and After Injury

3.2.1 Extracting Axoplasm from Naïve and Injured Sciatic Nerves

1. After anesthetizing the rats, the sciatic nerves are exposed through an incision and are subjected to unilateral sciatic nerve crush at mid-thigh level.
2. Surgical incision is closed with suture and monitored everyday.
3. Sciatic nerves ipsilateral (injured) and contralateral (uninjured) to crush injury are collected at 1, 4, 7, and 14 days of post-injury (dpi), and processed as described in Subheadings 3.1.1 and 3.1.2 (*see* **Note 18**).
4. Pool at least six injured sciatic nerves per each time point to obtain enough RNA for downstream small RNA enrichment.

3.2.2 RNA Isolation and Small RNA Fractionation

Refer Subheadings 3.1.3 and 3.1.4.

3.2.3 RNA Quantification and Reverse Transcription

Refer Subheading 3.1.5.

3.2.4 Real-Time PCR

1. Refer Subheading 3.1.6.
2. Data analysis can be done in similar way (*see* **Note 18**).

4 Results and Discussion

Several similar methods to the mechanical squeezing method detailed above have been used for biochemical and molecular studies of the composition of axoplasm in invertebrates as well as vertebrates [10, 12, 13, 44, 45]. Although these previous studies showed relatively enriched axoplasm that was largely free of glial contamination by western blotting, transcriptomic characterization of RNA contents in nerve tissues often failed due to significant levels of contamination. Therefore, minimizing non-neuronal contamination in the axoplasm extraction from nerve tissues is of essential importance in understanding how altered levels of axon-specific RNAs contributes to neurophysiological changes in response to distinct forms of local activity including injury [46, 47]. Although we have outlined the manual squeezing method for extracting nerve segments detailed, a key issue for success in this methodology is finding the right amount of pressure force that you need to get axoplasmic preparations before extracting non-neuronal materials. Manual removal of non-axonal surrounding tissues without stretching nerve tissue in cold PBS reduced the frequency of glial contamination. Since axoplasm is extruded through both ends of the nerve segments, cutting the nerve into smaller segments before squeezing helped limiting the number of pressure application step for axoplasm extraction.

As extended cycle PCR has proven rigorous for detection of low level of contamination in axonal preparations from cultured neurons, we assessed the purity of axoplasmic RNA preparations by extended RT-PCR method to validate the absence of neuronal cell body restricted mRNA (MAP2) and glial cells (GFAP, ErbB3, and H1F0). However, this is just likely the starting point from where researchers would add any control transcripts based on their rationale and interest.

We deliver qRT-PCR method detailed here as a means to assess if particular small RNAs of interest are present in the axon and whether their axonal levels are altered by extracellular stimuli. It can be also used for validation of high-throughput analytical approaches such as RNA deep sequencing, not only because qRT-PCR provides better estimates of expression levels of a particular small RNA through specific primers utilized, but also because individual genes of interest that have been identified from the deep sequencing need to be confirmed before downstream experiments. However, qRT-PCR can be very sensitive to the choice of reference genes.

Since any variation in the reference genes in samples will obviously obscure real changes in genes of interest and even result in data skewed significantly, it is critical to include an invariant reference gene(s) for data normalization in gene quantification analysis (*see* **Note 19**). Several different strategies have been introduced for normalizing qRT-PCR data including normalization to total cellular RNA content or an unregulated internal housekeeping gene. This may appear to be straightforward and applicable to small RNAs, but there is no consensus on reference gene for miRNA qRT-PCR. Some small RNA molecules, e.g., U6, 18S, tRNA or synthetic small RNAs as spike-in controls, have previously been used to normalize miRNA levels [48, 49], but these reference genes must be experimentally validated for specific experimental design. Mestdagh et al. [50] have showed that the man expression value of all tested small RNAs in a given sample serves better as a normalization control than the current strategy based on the use of a single or a set of reference genes. However, this method is not applicable for a small scale profiling study focusing only on a selected set of RNAs in functional studies. Therefore, we propose to identify a set of stable small RNA controls by running a pilot experiment in which small RNA controls can be identified. It is of utmost importance to use the validated reference genes for studying the biological significance of subtle changes in expression levels of RNA contents, and for confirmatory analyses of microarray/deep sequencing results by a means of qRT-PCR.

5 Notes

1. Four sciatic nerves were pooled together to obtain sufficient axonal RNA for downstream small RNA sequencing. To directly contrast RNA levels in axons and those in cell bodies, DRG can be collected and processed in parallel. For later processing, collect two sciatic nerves in a tube and store at −80 °C until axoplasm extraction to minimize freeze–thaw cycle. All manipulations for RNA isolation described here should be performed under RNase-free condition. Wipe out working area, pipettes and all the necessary equipment with RNase decontamination solution. Always use clean disposable gloves while handling all materials. Use nuclease-free barrier pipette tips to avoid RNase contamination.
2. Grab one edge of the nerve using a pair of forceps and carefully remove connective tissues on the nerve using the other pair of

fine forceps. Damage on nerve or epineurium during the process can cause RNA loss and non-neuronal cell contamination. Do not stretch the nerve extensively and grab the nerve only by the edge to avoid physical damage.

3. QIAzol lysis reagent from miRNsasy kit can be alternatively used for small RNA enrichment. Extended incubation of the nerve with TRIzol reagent causes nerve shrinkage and release of glial contaminants. Thus, minimizing processing time is critical for obtaining pure axoplasmic RNA. Try small volume of TRIzol to reduce chance of floating because the nerve segment is very slippery when it floats in the solution.
4. Do not rotate the pestle during squeezing.
5. Eight sciatic nerve segments (usually two dissected sciatic nerves) were processed in the same tube containing 0.2 mL of TRIzol on ice.
6. The reader who is not familiar with the whole procedure is recommended to process 3-4 segments at a time in a separate tube and pool RNA samples after confirming the purity of axoplasmic RNA.
7. For the initial experiment, it is recommended to check RNA integrity with total RNA using Bioanalyzer (Agilent). RNA integrity is critical especially for downstream RNA sequencing. It is possible to divide the sample into two volumes, one for total RNA and the other for small RNA fractionation.
8. 100 % ice-cold ethanol can be used, instead of isopropyl alcohol, to precipitate RNA by storing at −80 °C overnight.
9. Maximum loading volume for RNeasy Mini spin column and RNeasy MinElute spin column is ~0.7 mL. Repeat centrifugation on the same column until pass-through all sample.
10. Large RNA is retained inside the RNeasy Mini spin column and small RNA ≤200 nt is separated into the flow-through. Large RNA can be processed and used for axonal purity check (*see* **Note 14**) and comparative mRNA analysis parallel to small RNA profile. To isolate large RNA fraction from the RNeasy Mini spin column, save the RNeasy Mini spin column from Step 2 (Subheading 3.1.4) and wash with 0.7 mL Buffer RWT by centrifuge at ≥8000 × *g* for 15 s at room temperature. Then, follow steps 7–10 (Subheading 3.1.4) to purify large RNAs.
11. Since the Quant-iT RiboGreen reagent in DMSO may be absorbed into glass surfaces, it is advised to make the working solution in a disposable polypropylene plastic tube and protect from light by wrapping with foil.
12. To serve as an effective control, the RNA standard solution should be treated the same way as the experimental samples.

13. To minimize photobleaching effects, keep the time for fluorescence measurement constant for all samples. In case your value is saturated or below the lowest concentration of standard, repeats measure after reconstitution of RNA sample by increasing or decreasing the dilution ratio, respectively.
14. Axonal RNA extracted from sciatic nerve shows robust amplification of β-actin mRNA but should be devoid from MAP2 (microtubule-associated protein 2) mRNA which is known to be restricted in cell body. The glial specific mRNAs such as ErbB3 (Receptor tyrosine-protein kinase ErbB family-3), H1F0 (H1 histone family member 0) or GFAP (glial fibrillary acidic protein) can be used for examining glial cell contamination. Set up conventional PCR reaction with 0.5–1 ng of RNA. Run an agarose gel and confirm enrichment of β-actin PCR product.
15. NcodeVILO cDNA synthesis kit allows you to use universal reverse primer but forward primer should be designed specific to your target small RNA. Please read manufacturer's instruction to adjust Tm to reverse primer and to design specific primers. Run on validation experiment to ensure that the efficiencies of the target and endogenous control amplification are approximately equal.
16. Melting curve analysis at 60–95 °C (refer to instrument manual for specific programming) can be done to verify primer specificity. We confirmed a single peak in the melting curve analysis as well as a single band on agarose gels.
17. Review data to ensure that all reaction show amplification curves as expected (No template control should show negligible amplification). Threshold for cycling time (C_t) calculation can be manually set for all detectors to 0.2 alternatively (Threshold should locate within lower half of linear range in amplification curve and all the primers need to have similar efficiency).
18. Quantify the relative levels of individual transcripts by normalizing to an endogenous control (ΔC_t). Determine the $\Delta\Delta C_t$ value by subtracting the ΔC_t of injured axoplasm from ΔC_t of naïve axoplasm ($\Delta\Delta C_t = \Delta C_{t,\text{ injured axon}} - \Delta C_{t,\text{ naïve axon}}$). Express the fold difference as $2^{-\Delta\Delta Ct}$ which means relative amount of target small RNA in regenerating injured axon to that in naïve axon.
19. Our variance analysis in expression levels of each of the reference RNAs showed that U6 snRNA was the most consistently expressed RNA, followed by tRNA, 12 and 5S RNA. You can also try commonly expressed small RNA or mean of invariant small RNAs as reference RNA which is robustly expressed in the sample.

Acknowledgements

The methods presented here were developed using funds from the National Institutes of Health (P20-GM103464 and R21-NS085691 to SY). This work was also partially supported by Delaware INBRE Core Center Access Award (SY) from an Institutional Development Award (IDeA) Network of Biomedical Research Excellence program (INBRE; P20-GM103446) of the National Institutes of Health.

References

1. Eva R et al (2012) Intrinsic mechanisms regulating axon regeneration: an integrin perspective. Int Rev Neurobiol 106:75–104
2. Ferreira LM et al (2012) Neural regeneration: lessons from regenerating and non-regenerating systems. Mol Neurobiol 46 (2):227–241
3. Pernet V, Schwab ME (2014) Lost in the jungle: new hurdles for optic nerve axon regeneration. Trends Neurosci 37(7):381–387
4. Snider WD et al (2002) Signaling the pathway to regeneration. Neuron 35(1):13–16
5. Chen ZL et al (2007) Peripheral regeneration. Annu Rev Neurosci 30:209–233
6. Baleriola J, Hengst U (2015) Targeting axonal protein synthesis in neuroregeneration and degeneration. Neurotherapeutics 12(1):57–65
7. Wu D, Murashov AK (2013) Molecular mechanisms of peripheral nerve regeneration: emerging roles of microRNAs. Front Physiol 4:55
8. Yoo S et al (2010) Dynamics of axonal mRNA transport and implications for peripheral nerve regeneration. Exp Neurol 223(1):19–27
9. Ben-Yaakov K et al (2012) Axonal transcription factors signal retrogradely in lesioned peripheral nerve. EMBO J 31(6):1350–1363
10. Hanz S et al (2003) Axoplasmic importins enable retrograde injury signaling in lesioned nerve. Neuron 40(6):1095–1104
11. Perry RB et al (2012) Subcellular knockout of importin beta1 perturbs axonal retrograde signaling. Neuron 75(2):294–305
12. Yudin D et al (2008) Localized regulation of axonal RanGTPase controls retrograde injury signaling in peripheral nerve. Neuron 59 (2):241–252
13. Perlson E et al (2005) Vimentin-dependent spatial translocation of an activated MAP kinase in injured nerve. Neuron 45(5):715–726
14. Verma P et al (2005) Axonal protein synthesis and degradation are necessary for efficient growth cone regeneration. J Neurosci 25 (2):331–342
15. Donnelly CJ et al (2011) Limited availability of ZBP1 restricts axonal mRNA localization and nerve regeneration capacity. EMBO J 30 (22):4665–4677
16. Donnelly CJ et al (2013) Axonally synthesized beta-actin and GAP-43 proteins support distinct modes of axonal growth. J Neurosci 33 (8):3311–3322
17. Yoo S et al (2013) A HuD-ZBP1 ribonucleoprotein complex localizes GAP-43 mRNA into axons through its 3′ untranslated region AU-rich regulatory element. J Neurochem 126 (6):792–804
18. Merianda TT et al (2015) Axonal amphoterin mRNA is regulated by translational control and enhances axon outgrowth. J Neurosci 35 (14):5693–5706
19. Merianda TT et al (2013) Axonal transport of neural membrane protein 35 mRNA increases axon growth. J Cell Sci 126(Pt 1):90–102
20. Martin KC, Ephrussi A (2009) mRNA localization: gene expression in the spatial dimension. Cell 136(4):719–730
21. Besse F, Ephrussi A (2008) Translational control of localized mRNAs: restricting protein synthesis in space and time. Nat Rev Mol Cell Biol 9(12):971–980
22. Jung H et al (2014) Remote control of gene function by local translation. Cell 157 (1):26–40
23. Jung H et al (2012) Axonal mRNA localization and local protein synthesis in nervous system assembly, maintenance and repair. Nat Rev Neurosci 13(5):308–324
24. Bushati N, Cohen SM (2007) microRNA functions. Annu Rev Cell Dev Biol 23:175–205
25. Schratt G (2009) Fine-tuning neural gene expression with microRNAs. Curr Opin Neurobiol 19(2):213–219

26. Liu NK, Xu XM (2011) MicroRNA in central nervous system trauma and degenerative disorders. Physiol Genomics 43(10): 571–580
27. Ishizu H et al (2012) Biology of PIWI-interacting RNAs: new insights into biogenesis and function inside and outside of germlines. Genes Dev 26(21):2361–2373
28. Bartel DP (2009) MicroRNAs: target recognition and regulatory functions. Cell 136 (2):215–233
29. Kosik KS (2006) The neuronal microRNA system. Nat Rev Neurosci 7(12):911–920
30. Kye MJ et al (2014) SMN regulates axonal local translation via miR-183/mTOR pathway. Hum Mol Genet 23(23):6318–6331
31. Kar AN et al (2013) Intra-axonal synthesis of eukaryotic translation initiation factors regulates local protein synthesis and axon growth in rat sympathetic neurons. J Neurosci 33 (17):7165–7174
32. Liu CM et al (2013) MicroRNA-138 and SIRT1 form a mutual negative feedback loop to regulate mammalian axon regeneration. Genes Dev 27(13):1473–1483
33. Zhang Y et al (2013) The MicroRNA-17-92 cluster enhances axonal outgrowth in embryonic cortical neurons. J Neurosci 33 (16):6885–6894
34. Taylor AM et al (2009) Axonal mRNA in uninjured and regenerating cortical mammalian axons. J Neurosci 29(15):4697–4707
35. Zivraj KH et al (2010) Subcellular profiling reveals distinct and developmentally regulated repertoire of growth cone mRNAs. J Neurosci 30(46):15464–15478
36. Gumy LF et al (2011) Transcriptome analysis of embryonic and adult sensory axons reveals changes in mRNA repertoire localization. RNA 17(1):85–98
37. Natera-Naranjo O et al (2010) Identification and quantitative analyses of microRNAs located in the distal axons of sympathetic neurons. RNA 16(8):1516–1529
38. Sasaki Y et al (2014) Identification of axon-enriched microRNAs localized to growth cones of cortical neurons. Dev Neurobiol 74 (3):397–406
39. Wu D et al (2011) MicroRNA machinery responds to peripheral nerve lesion in an injury-regulated pattern. Neuroscience 190:386–397
40. Yu B et al (2011) Altered microRNA expression following sciatic nerve resection in dorsal root ganglia of rats. Acta Biochim Biophys Sin (Shanghai) 43(11):909–915
41. Dajas-Bailador F et al (2012) microRNA-9 regulates axon extension and branching by targeting Map1b in mouse cortical neurons. Nat Neurosci 15:697–699
42. Phay M et al (2015) Dynamic change and target prediction of axon-specific microRNAs in regenerating sciatic nerve. PLoS One 10(9): e0137461
43. Yu B et al (2011) Profile of microRNAs following rat sciatic nerve injury by deep sequencing: implication for mechanisms of nerve regeneration. PLoS One 6(9):e24612
44. Brady ST et al (1985) Video microscopy of fast axonal transport in extruded axoplasm: a new model for study of molecular mechanisms. Cell Motil 5(2):81–101
45. Schmied R et al (1993) Endogenous axoplasmic proteins and proteins containing nuclear localization signal sequences use the retrograde axonal transport/nuclear import pathway in Aplysia neurons. J Neurosci 13(9):4064–4071
46. Perlson E et al (2004) Differential proteomics reveals multiple components in retrogradely transported axoplasm after nerve injury. Mol Cell Proteomics 3(5):510–520
47. Rishal I et al (2010) Axoplasm isolation from peripheral nerve. Dev Neurobiol 70 (2):126–133
48. Kanemaru H et al (2011) The circulating microRNA-221 level in patients with malignant melanoma as a new tumor marker. J Dermatol Sci 61(3):187–193
49. Liu R et al (2011) A five-microRNA signature identified from genome-wide serum microRNA expression profiling serves as a fingerprint for gastric cancer diagnosis. Eur J Cancer 47 (5):784–791
50. Mestdagh P et al (2009) A novel and universal method for microRNA RT-qPCR data normalization. Genome Biol 10(6):R64

Neuromethods (2017) 128: 161–181
DOI 10.1007/7657_2016_5

Published online: 31 July 2016

In Ovo Electroporation of miRNA Plasmids to Silence Genes in a Temporally and Spatially Controlled Manner

Nicole H. Wilson and Esther T. Stoeckli

Abstract

The ability to spatially and temporally control gene expression during development is crucial for the elucidation of gene function in vivo. The use of RNA interference (RNAi)-based technologies in combination with oviparous animal models allows for efficient, precise gene silencing. We have developed approaches using RNAi in the chicken embryo to analyze gene function during neural tube development. Here we describe the construction of plasmids that direct the expression of one or two artificial microRNAs (miRNAs) to knock down endogenous target protein/s upon electroporation into the spinal cord. The miRNA cassette is directly linked to a fluorescent protein reporter, which allows the faithful visualization of transfected cells. Different promoters/enhancers drive transcript expression in genetically defined cell subpopulations in the neural tube. Mixing multiple RNAi vectors allows combinatorial knockdowns of two or more genes in different cell types, thus permitting the rapid analysis of complex cellular and molecular interactions.

Keywords: Neural development, Gene knockdown, Spinal cord, Commissural neuron, Floor plate, MicroRNA, Axon guidance, RNA interference, RNAi, Chicken embryo

1 Introduction

Gene function during development is cell type specific and context dependent, and as such the function of novel genes should ideally be examined in the context of the whole organism. Many genes are active during multiple time windows, which adds additional complexity to the experimental approach. Classical gene knockouts in the mouse, for example, may cause early embryonic lethality that precludes the analysis of gene function at later stages of development. Thus, the ability to perturb gene function in a spatially and temporally controlled manner is extremely advantageous. While this can be achieved to some extent by creating conditional knockout mice [1], this is a laborious and expensive undertaking that is rate limiting in the elucidation of gene function.

Because of its easy accessibility in the egg, the chicken embryo has been a popular model for developmental in vivo studies. In recent years, RNA interference (RNAi) was developed as a powerful method to perturb gene expression during development.

Double-stranded RNA in the form of short-interfering RNA (siRNA), long double-stranded RNA (dsRNA), short-hairpin RNA (shRNA), or microRNA (miRNA) can be delivered to cells by applying an electric field (electroporation) [2–9]. These approaches harness the naturally occurring miRNA biogenesis pathway, a posttranscriptional gene regulatory mechanism that evokes gene silencing by activating the RNA-induced silencing complex (RISC) (reviewed in [10]). By choosing the time point and positioning the electrodes appropriately, different cell populations can be selected for gene silencing in a spatiotemporally controlled manner. However, many of these approaches lack cell type specificity, which can be problematic for the following reasons: (1) manual positioning of the electrodes does not allow targeting of distinct subpopulations of cells located in the same physical area; (2) genes located in disparate subpopulations of cells are difficult to downregulate simultaneously; and (3) identification of cells experiencing gene knockdown is often indirect (for example, by the co-electroporation of a fluorescent reporter). We recently developed an improvement on current techniques to markedly enhance the precision and combinatorial possibilities of in ovo RNAi, using a miRNA-based approach [11]. In the protocol below, we explain how to clone and apply these miRNA constructs in vivo.

In each construct, one or two artificial miRNAs are directly coupled to the open reading frame encoding a fluorescent protein in a single transcript that allows for direct visualization of transfected cells. Cell type-specific promoters/enhancers drive transcript expression. Electroporation of these constructs into the developing chick neural tube enables effective knockdown of target genes in different cell types, which can be traced directly by the expression of fluorescent markers. These cell type-specific miRNA constructs can also be applied in combination, opening a wide range of possibilities to study complex molecular interactions in vivo. While we have concentrated on the applicability of these constructs to the study of commissural axon guidance in the spinal cord, the use of other cell type-specific promoters/enhancers would allow the easy adaptation of the constructs to diverse subsets of neurons and their targets in the neural tube (for example, in motor neurons [HB9; [12]], primary sensory neurons [CREST3; [13]], subpopulations of dorsal interneurons [14], and other cell types [15]).

In the protocol below, we explain how to design and construct the miRNA-based plasmids by PCR and molecular cloning, and then how to electroporate them into the chick neural tube to silence genes in a spatially and temporally controlled manner. We also discuss methods for assessing gene downregulation and the phenotypic effects of gene silencing.

2 Materials

2.1 Construction of miRNA Plasmids

1. Backbone vectors: Plasmids containing a cell type-specific promoter/enhancer, a fluorescent protein marker, and an artificial miRNA expression cassette (excised from pRFPRNAi [9]) were constructed as described [11] (Fig. 1). The chicken β-actin promoter is used to drive ubiquitous expression, while an enhancer of mouse Hoxa1 drives expression specifically in the floor plate and an enhancer of mouse Atonal homolog 1 (Math1) drives expression specifically in commissural dI1 neurons of the spinal cord. The fluorescent proteins used were monomeric red fluorescent protein 1 (mRFP1; Clontech), enhanced blue fluorescent protein 2 (EBFP2; Addgene), humanized *Renilla* green fluorescent protein II (hrGFPII; Stratagene), enhanced green fluorescent protein (EGFP; Clontech), farnesylated (membrane-bound) tandem-dimer Tomato red fluorescent protein (tdTomatoF; Clontech), and farnesylated EGFP (EGFPF) (*see* **Note 1**).

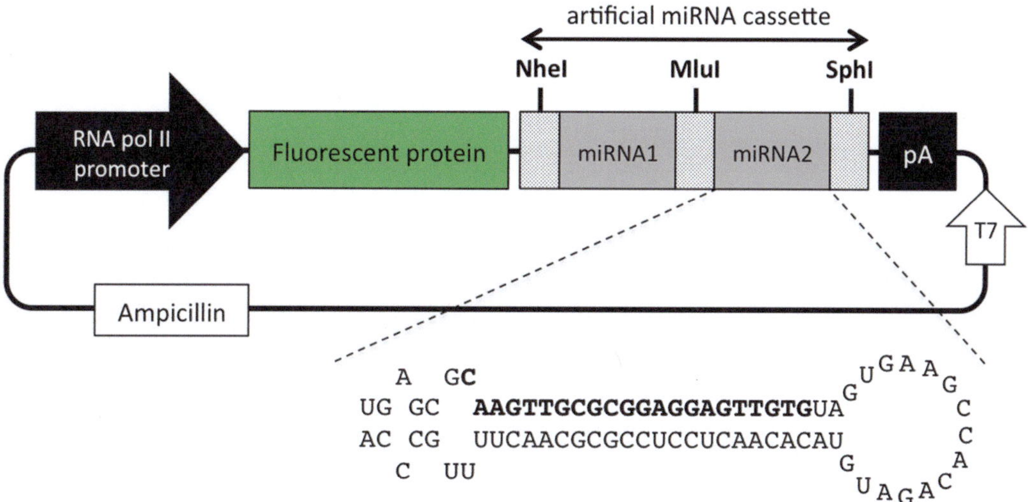

Fig. 1 Schematic of miRNA-based constructs. Each construct contains a promoter/enhancer driven by RNA polymerase II, which determines the cell type in which the RNAi construct is expressed. The β-actin promoter drives ubiquitous expression in the neural tube, while the Math1 enhancer (with minimal β-globin promoter) drives expression in commissural dI1 neurons, and the Hoxa1 enhancer (with minimal TATA box promoter) drives specific expression in the floor plate (*see* text). The various promoters produce a single transcript encoding a fluorescent protein directly coupled to an artificial miRNA cassette, located within the 3′UTR. There is a poly-A tail (pA) at the 3′ end. The artificial miRNA cassette contains insertion sites for one or two artificial miRNAs (miRNA1, miRNA2). These are flanked by unstructured sequences containing unique restriction sites (NheI, MluI, SphI) used for cloning the miRNAs into the vector. The stem-loop structure of the miRNA30-like cassette is shown for a control artificial miRNA against firefly Luciferase (miLuc). The target (sense) sequence of Luciferase is indicated in bold. A T7 promoter (T7) enables sequencing of the miRNA cassette in the 3′–5′ direction. The construct encodes for ampicillin resistance

2. DNA polymerase: 3 U/μl Pfu and 10× Pfu reaction buffer.
3. Nucleotides: 10 mM mix of PCR-grade dNTPs.
4. DNA purification kit: For example, Wizard® SV Gel and PCR Cleanup System (Promega).
5. Restriction enzymes and 10× reaction buffers (NEB): NheI, MluI, SphI.
6. TBE buffer: 5.4 g Tris base, 5.5 g boric acid, 2 ml 0.5 M EDTA in ddH_2O (double-distilled water), 1 l final volume.
7. 10× DNA loading buffer: 3.9 ml glycerol, 500 μl 10 % (w/v) SDS, 200 μl 0.5 M EDTA, 0.025 g bromophenol blue, and 0.025 g xylene cyanol, in ddH_2O to a final volume of 10 ml.
8. Agarose gel: 1 % DNA-grade agarose (e.g., peqGOLD Universal Agarose) in TBE buffer. Heat carefully in a microwave oven to dissolve the agarose. Add ethidium bromide to a final concentration of 0.5 μg/ml. Allow to cool to approximately 60 °C before pouring the gel into a casting tray.
9. DNA ligase: T4 DNA ligase and 10× ligase reaction buffer.
10. Competent cells (e.g., XL1-blue, DH5α).
11. LB broth: Add 10 g Bacto-tryptone, 5 g yeast extract, and 10 g NaCl to 900 ml ddH_2O, adjust to pH 7.5 with NaOH, add ddH_2O to a final volume of 1 l. Autoclave.
12. LB agar: Add 10 g Bacto-tryptone, 5 g yeast extract, 10 g NaCl, and 15 g agar to 900 ml ddH_2O, adjust to pH 7.5, adjust to a final volume of 1 l. Autoclave.
13. Ampicillin: Supplement LB broth and LB agar plates with 100 μg/ml ampicillin before use.
14. Plasmid mini- and maxi-prep kits (e.g., Nucleobond®, Nucleospin® kits; Machery-Nagel).
15. T7 primer (10 μM), for sequencing: 5′-TAATACGACTCA CTATAGGG.
16. 10× Sequencing buffer: 100 mM Tris (pH 8.0), 0.1 mM EDTA.

2.2 Windowing Eggs

1. Fertilized chicken eggs, from a local hatchery.
2. Incubator set at 38.5 °C and 45 % humidity (e.g., Juppiter 576 Setter + Hatcher; FIEM, Italy, or Heraeus/Kendro Model B12, Kendro Laboratory Products, Germany).
3. Heating plate set at 80 °C, to melt paraffin.
4. Paraffin wax (Paraplast tissue-embedding medium) in a glass beaker.
5. 70 % ethanol in a spritz bottle.
6. Facial tissues.

7. Scotch tape.
8. Scalpel with elongated triangular blade.
9. Sterile syringe with 18G needle.
10. Paintbrush.
11. Fine dissection scissors (e.g., Fine Science Tools; 91460-11).

2.3 In Ovo Electroporation

1. Borosilicate glass capillaries: outer Ø/inner Ø: 1.2 mm/ 0.68 mm (World Precision Instruments; 1B120F-4).
2. Glass needle puller (e.g., Narishige PC-10).
3. Square wave electroporator (e.g., BTX ECM 830).
4. Spring scissors (Fine Science Tools; 15003-08).
5. Dumont #5 forceps (Fine Science Tools; 11252-20).
6. Platinum electrodes: 4 mm length, 4 mm distance between cathode and anode.
7. Polyethylene tubing: Ø 1.24 mm.
8. 0.2 μm Syringe-driven filter units.
9. Trypan blue solution (Gibco, Life Technologies), 0.4 %, filtered.
10. 20× Sterile phosphate-buffered saline (PBS): 160 g of NaCl, 4 g of KCl, 28.8 g of $Na_2HPO_4 \cdot 2\ H_2O$, 4.1 g of $NaH_2PO_4 \cdot H_2O$ in 1 l ddH_2O. Autoclave.
11. PBS: Add 50 ml of 20× PBS to 950 ml of ddH_2O. Autoclave. Store in 20–50 ml aliquots, taking a fresh tube for each day of electroporation.
12. Transfer pipettes.
13. Wash bottle filled with ddH_2O.

3 Methods

3.1 Generation of miRNA-Expressing Plasmids

The artificial miRNAs are generated by PCR. It is recommended to first construct and screen 3–6 miRNAs per gene to identify those that effectively knock down the gene of interest, using in situ hybridization, immunohistochemistry, Western blots, or in vitro assays (*see* Section 3.4). Effective miRNAs should also be screened for off-target effects (*see* **Note 2**).

1. *Design PCR primers:* Cloning into the first and second miRNA insertion sites requires distinct sets of universal and gene-specific primers. The universal primers are the following:
   (a) For the first hairpin site:
   5′ primer1:

5′-GGCGGGGCTAGCTGGAGAAGATGCCTTCCG-GAGA GGTGCTGCTGAGCG
3′ primer1:
5′-GGGTGGACGCGTAAGAGGGGAAGAAAGCTTC-TAA CCCCGCTATTCACCACCACTAGGCA

(b) For the second hairpin site:

5′ primer2:
5′-GGCGGGACGCGTGCTGTGAAGATCCGAA-GATGCCTT GCGCTGGTTCCT CCGTGAGCG
3′ primer2:
5′-CGCCGCGCATGCACCAAGCAGAGCAGCCTGA AGACCAGTAGGCA

The gene-specific primers encoding the artificial miRNAs are designed using the siRNA Target Finder from Genscript: https://www.genscript.com/ssl-bin/app/rnai (*see* **Note 3**). We usually use the Statistical Model (Pattern: AAN19). Enter the appropriate information, as directed, to generate a list of 21 bp candidate target sequences. Select 3–6 of these and incorporate them into primers as shown below. The primers include miRNA-flanking sequences and common stem/loop sequences (from human miRNA30) [9].

In the examples below, the gene-specific target sequences are underlined. Note that there is a mismatch at the 5′ base of the forward strand (shown in bold) to mimic the natural mismatch in miRNA30 at this position.

(c) Primers for cloning a gene-specific miRNA into the first hairpin site (HP1):

Example target sequence (21 nt, derived from Contactin2 (Cntn2)):
5′-AAGGCACTTATGAGTGCGAGG
Cntn2 forward HP1 = 58mer:
5′-GAGAGGTGCTGCTGAGCG<u>**C**AGGCACTTA TGAGTGCGAGG</u>TAGTGAAGCCACAGATGTA
Cntn2 reverse HP1 = 57mer
5′-ATTCACCACCACTAGGCA<u>AAGGCA CTTATGAGTGCGAGG</u>TACATCTGTGGCTTCACT

(d) Primers for cloning a gene-specific miRNA into the second hairpin site (HP2):
Example target sequence (21 nt, derived from firefly Luciferase (Luc)):
5′-AAAGTTGCGCGGAGGAGTTGTG
Luc forward HP2:
5′-CTGGTTCCTCCGTGAGCG<u>**C**AAGTTGCGC GGAGGAGTTGTG</u>TAGTGAAGCCACAGATGTA
Luc reverse HP2:

5′-CCTGAAGACCAGTAGGCAAAAGTTGCGCG
GAGGAGTTGTGTACATCTGTGGCTTCACT

2. *Perform PCR reaction:* The miRNA30-like hairpins (with chicken miRNA flanking sequences) are generated in a PCR reaction using the universal primers together with the gene-specific primers. The PCR reaction is set up in a 50 μl reaction volume as follows:

   (a) Cloning into first hairpin site:

   1 μl (10 ng) Cntn2 forward HP1
   1 μl (10 ng) Cntn2 reverse HP1
   1 μl (100 ng) 5′ primer HP1
   1 μl (100 ng) 3′ primer HP1
   1 μl dNTPs (10 mM)
   5 μl 10× Pfu reaction buffer
   1 μl Pfu DNA polymerase
   39 μl PCR-grade water

   (b) Cloning into second hairpin site:

   1 μl (10 ng) Luc forward HP2
   1 μl (10 ng) Luc reverse HP2
   1 μl (100 ng) 5′ primer HP2
   1 μl (100 ng) 3′ primer HP2
   1 μl dNTPs (10 mM)
   5 μl 10× Pfu reaction buffer
   1 μl Pfu DNA polymerase
   39 μl PCR-grade water

   PCR cycling conditions are as follows: 94 °C for 1 min; 30 cycles of 94 °C for 30 s, 55 °C for 30 s, and 72 °C for 1 min; and 72 °C for 9 mins.

3. Purify the PCR product using the Wizard® SV Gel and PCR Cleanup System, following the manufacturer's instructions. Collect the purified PCR product in 30 μl ddH_2O. Digest both the PCR product and 1 μg of the appropriate backbone vector with restriction enzymes (*see* Fig. 1):

   (a) First hairpin site: use NheI and MluI
   (b) Second hairpin site: use MluI and SphI

4. Use TBE agarose electrophoresis (1 % agarose) to verify the success of the PCR and digests. The PCR product should be approximately 160 bp long. Excise and purify the DNA bands using the Wizard® SV Gel and PCR Cleanup System. Elute the purified DNA in 15 μl of ddH_2O.

5. Set up the ligation reaction as follows: add 5 μl miRNA insert DNA (digested PCR product), 3 μl digested vector DNA, 1 μl 10× T4 DNA ligase buffer, and 1 μl T4 DNA ligase (3 U/μl)

to a tube. Mix gently and incubate at room temperature for 2 h or at 4 °C overnight.

6. Follow standard techniques for heat-shock transformation of competent bacterial cells with 3 μl of the ligation mix. Plate cells on LB agar (containing ampicillin) and harvest DNA (4–6 colonies from each ligation reaction) by plasmid minipreparation (e.g., Nucleospin® Plasmid kit, Machery-Nagel).
7. *Sequencing miRNA plasmids:* Sanger sequencing using a T7 primer should reveal the sequence of the poly-A tail and miRNA cassette in the 3′–5′ direction (*see* Fig. 1). Under standard conditions the sequencing reaction often fails due to strong secondary hairpin structure of the miRNAs. To help improve sequencing, perform the following steps to increase the conversion of supercoiled DNA to ssDNA, which is more amenable to sequencing [16]:
   (a) Perform sequencing reaction in 10 mM Tris–Cl with 0.01 mM EDTA (pH 8.0) instead of water (*see* **Note 4**).
   (b) Add a heat denaturation step (98 °C, 5 min in a PCR machine) prior to sequencing.
8. Grow sequence-verified plasmids as midi- or maxi-preparations (e.g., Nucleobond® Xtra Midi kit, Machery-Nagel). Suspend plasmid DNA in sterile ddH_2O (*see* **Note 5**), measure concentration by spectrophotometry (*see* **Note 6**), and store at −20 °C.
9. *DNA injection mix:* The appropriate DNA concentration must be determined by the user and will vary according to the enhancer/promoter used to drive expression. As a guideline, we typically use 0.2–1.0 μg/μl (*see* **Note 7**). In a total of 20 μl (in sterile ddH_2O), the injection mixture should contain X μl miRNA plasmid DNA, 1 μl 20× PBS, and 2 μl 0.4 % trypan blue.

3.2 Windowing Eggs

Access to the developing embryo is obtained by cutting a window in the eggshell (Fig. 2). A detailed description can also be found as a video online [17]:

1. Incubate eggs at 38.5 °C and 45 % humidity (*see* **Notes 8–10**) until the embryos have reached the desired stage for experimental manipulation (*see* **Note 11**). Stage embryos according to Hamburger and Hamilton [18]. For in ovo electroporation of the spinal cord, we usually use embryos on embryonic day 2 (E2; HH12–14) or E3 (HH17–18).
2. Place the egg with the long axis lying horizontally. Incubate for 20 min to allow the embryo to reposition on top of the egg yolk before windowing.

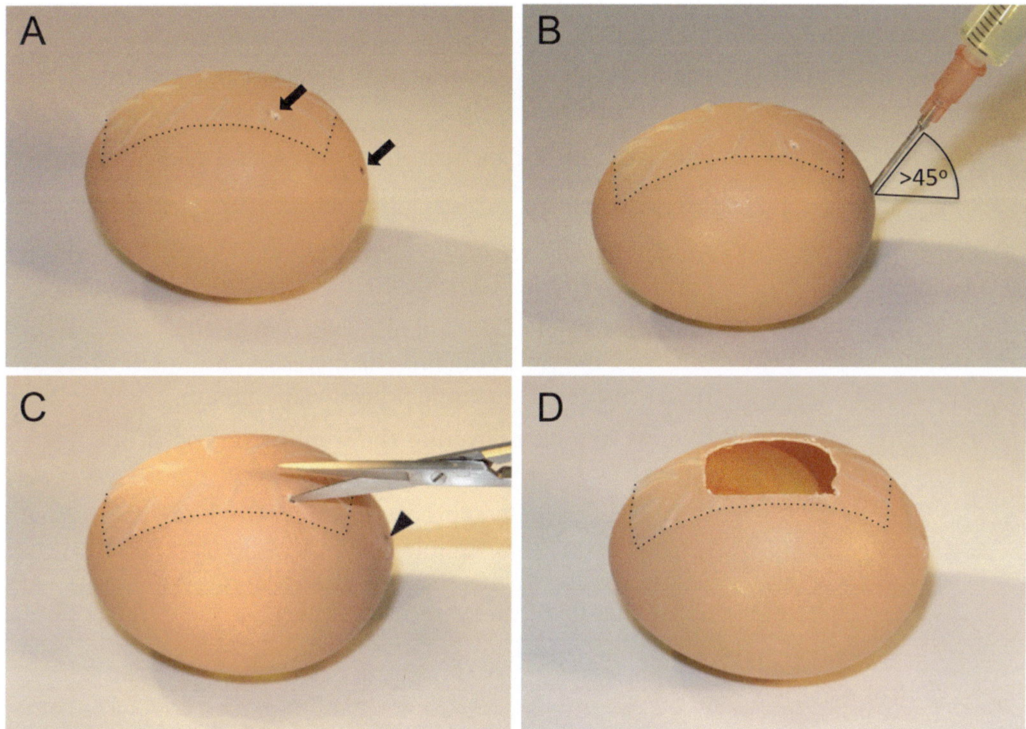

Fig. 2 Windowing eggs. (**a**) A piece of tape (indicated with *dotted lines*) is placed on the top surface of the egg. Two small holes are drilled in the eggshell (*arrows*), one at the blunt end of the egg, the other at the corner of the intended window. (**b**) A syringe with 18G needle is inserted into the hole at the blunt end of the egg at an angle of >45 °. Approximately 3 ml of albumen is withdrawn. (**c**) The hole at the blunt end is sealed with paraffin wax (*arrowhead*). Small dissection scissors are used to carefully cut a window (approximately 2 × 2 cm) in the eggshell. (**d**) Windowed egg. The embryo should be visible in the middle of the window

3. While embryos are repositioning, turn on the heating plate and melt the paraffin wax.
4. Wipe each egg (*see* **Note 12**) and your workspace with 70 % ethanol (*see* **Note 13**).
5. Stick Scotch tape along the long axis on top of the egg in the area of the intended window (Fig. 2a). This will prevent large cracks and pieces of eggshell from falling onto the embryo.
6. Carefully drill a small hole into the blunt end of the egg and into a corner of the intended window on top of the egg (Fig. 2a; *see* **Note 14**), using the pointed tip of the scalpel blade.
7. Push the 18G needle of the syringe into the hole at the blunt end of the egg and remove ~3 ml of albumen. To avoid damage to the egg yolk, insert the needle at an angle greater than 45 ° (Fig. 2b) (*see* **Note 15**).
8. Use a paintbrush to seal the hole at the blunt end (and any other cracks in the eggshell) with melted paraffin.

9. Using the hole on top of the egg as a starting point, cut a window into the eggshell, carefully holding scissors horizontally to avoid damaging the embryo (Fig. 2c, d; *see* **Note 16**).
10. Seal the window with Scotch tape (*see* **Note 17**) and put the egg back into the incubator at 38.5 °C and 45 % humidity (*see* **Notes 9** and **13**).

3.3 In Ovo Electroporation

1. Autoclave your tools and clean your working space with 70 % ethanol (*see* **Note 12**).
2. Pull glass capillaries to make injection needles and carefully break off the tip to obtain a diameter of ~5 μm (*see* **Note 18**). Insert the needle into the polyethylene tubing.
3. Remove the tape covering the window on top of the egg (*see* **Note 13**).
4. Place the egg in an appropriate holder (*see* **Note 19**) so that it is manoeuvrable but stable during the electroporation. If necessary, slightly rotate the egg until the embryo is positioned in the middle of the window (Fig. 3a).

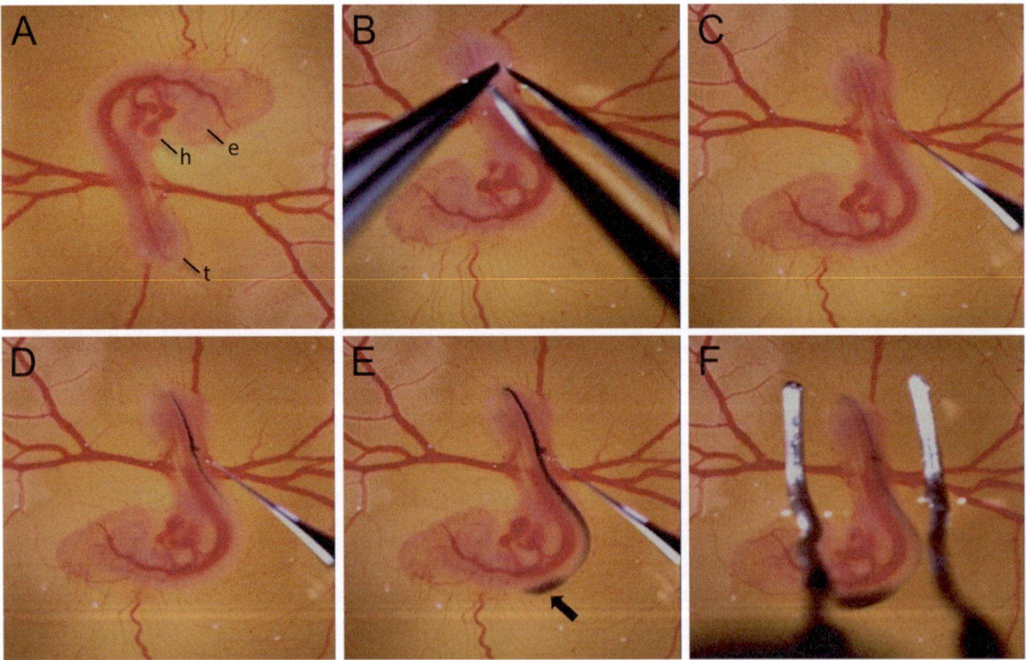

Fig. 3 In ovo electroporation. (**a**) Embryo in the egg. The eye (*e*), heart (*h*), and tail (*t*) are indicated. (**b**) For right-handed users, the embryo is rotated 180 ° to improve access to the neural tube (see Note 21). The vitelline membranes are removed by lifting with forceps and gently tearing towards the tail. (**c**) A thin glass needle containing the injection solution with trypan blue dye is inserted into the neural tube. (**d**) The solution is slowly injected. A thin line of blue dye visualizes the extent of injection within the neural tube. (**e**) When the dye reaches the level of the hindbrain (*arrow*), the needle is slowly withdrawn. Carefully avoid leakage of the injected solution at any time to get efficient transfection. (**f**) A few drops of PBS are applied to the embryo. The electrodes are positioned appropriately, and the electric field is applied

5. Remove the vitelline membranes using fine forceps and spring scissors to gain direct access to the embryo (Fig. 3b; *see* **Note 20**).
6. Inject the DNA solution (*see* Section 3.1, **step 9;** *see* **Notes 5–7**) into the central canal of the spinal cord at the lumbosacral level (Fig. 3c–e; *see* **Note 21**). Control injection volume by mouth, making sure that the solution does not leak (*see* **Note 22**).
7. Using a transfer pipette, add a few drops of sterile PBS to the embryo to prevent overheating and high electrical resistance during subsequent electroporation.
8. Place the electrodes parallel to the anterior-posterior axis of the spinal cord (Fig. 3f). In order to avoid bleeding, do not touch the blood vessels while applying current (*see* **Note 23**). Electroporate the embryo with five pulses of 50-ms duration at 18 V for E2, or at 25 V for E3 embryos with an interpulse interval of 1 s (*see* **Notes 24** and **25**).
9. Wash the electrodes with ddH_2O to remove accumulated protein (white deposits) immediately after electroporating the embryo.
10. Reseal the egg with tape (or alternatively, use a cover slip as previously described [19]; *see* **Note 26**) and put it back into the incubator until the embryo reaches the desired developmental stage for analysis.

3.4 Verification of Gene Knockdown

A crucial factor in the use of miRNA-based plasmids as functional tools in developmental biology is the selection of effective, specific miRNAs [10, 20]. This process should incorporate the identification of several independent miRNAs to allow any observed phenotype to be confirmed, negative controls (scrambled and/or unrelated miRNAs) and rescue constructs.

We have previously used COS-7 cells to pre-screen a number of novel candidate miRNAs in vitro, which can accelerate the miRNA selection procedure (Fig. 4a). The cells are repeatedly transfected with a miRNA construct, followed by transfection of its target gene and immunolabeling to assess the efficiency of gene knockdown [11]. Artificial miRNAs that are successful in vitro are then further tested in vivo to confirm their efficacy (*see* **Note 27**). A control miRNA (against firefly luciferase, for example) must be used for both the in vitro screening and in vivo electroporation [9, 11, 20].

All constructs must be efficiently electroporated in order to achieve strong downregulation. Handling chicken embryos in ovo requires manual skills that must be acquired through training. We and others have previously published troubleshooting tips for this procedure [17, 19, 21, 22]. An advantage of the current method is that the delivery of miRNA is directly visualized by the expression

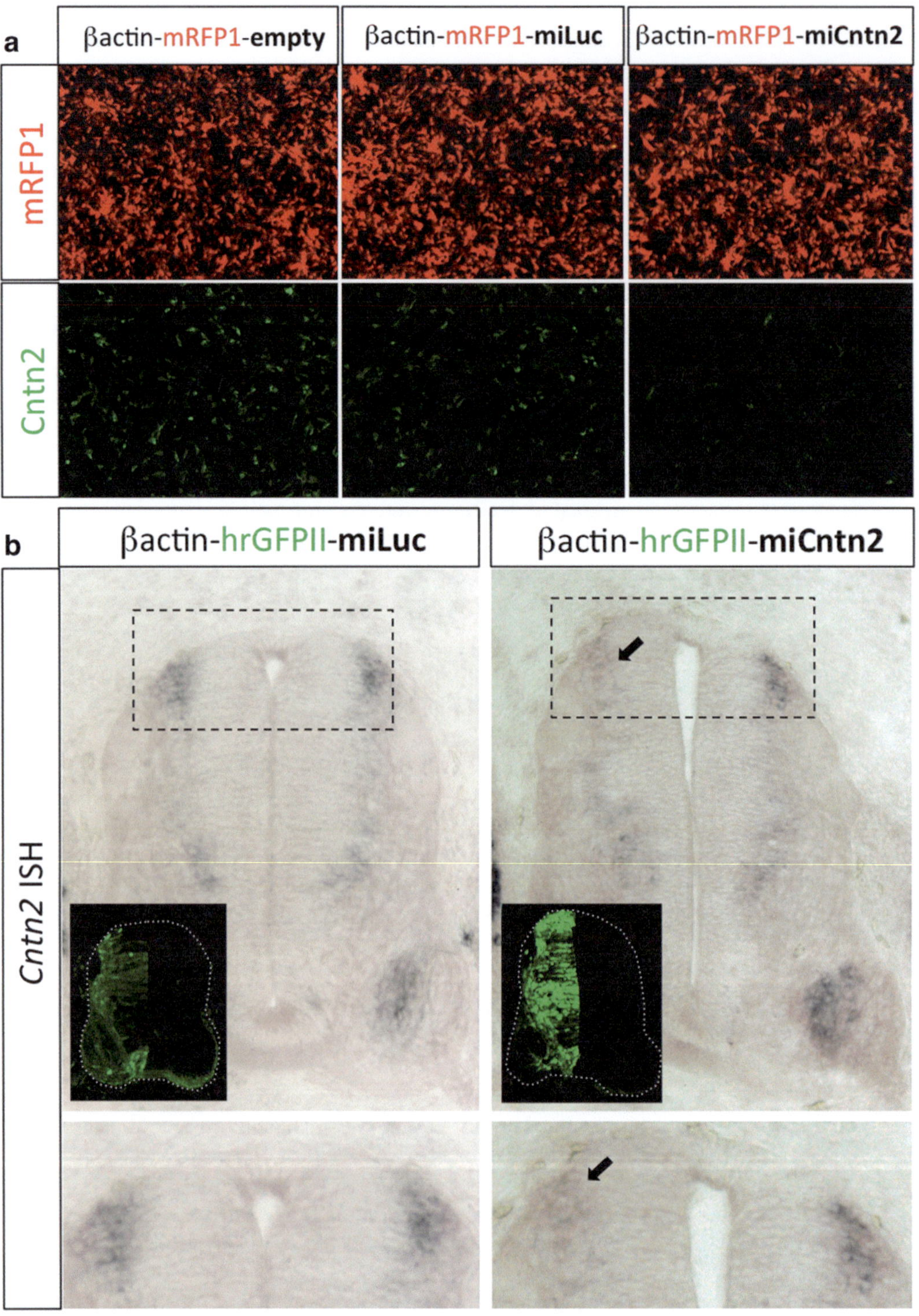

Fig. 4 Determining miRNA knockdown efficacy in vitro and in vivo. (**a**) Testing miRNAs in vitro. COS-7 cells were transfected with miRNA constructs twice over 24 h. In this example, the constructs used were β-actin-mRFP1 containing an empty miRNA cassette (β-actin-mRFP1-empty), miRNA against Luciferase

of a fluorescent protein encoded by the same construct as the miRNA. Electroporated embryos can be assessed under a fluorescent stereomicroscope, either directly in ovo or ex ovo in a petri dish. Those showing strong expression of fluorescent protein in the neural tube (indicating efficient electroporation) can be preselected for subsequent dissections and analyses (*see* **Note 28**).

There are several techniques to assess the efficiency of gene knockdown in vivo (*see* **Note 29**). These include:

1. Immunohistochemistry on cryosections [2, 3, 11, 23].
2. Western blot analysis of protein levels using lysates from electroporated spinal cord extracts [24].
3. In situ hybridization analysis on cryosections [3, 11, 25] (Fig. 4b).
4. RT-PCR to show downregulation of the targeted transcript [26].

In most cases, the electroporated hemi-segment of the spinal cord can be compared to the non-treated half, which serves as an internal control (Fig. 4b).

3.5 Combinatorial Gene Knockdowns

The miRNA plasmids described here contain insertion sites for two different miRNAs (Fig. 1). Their simultaneous activity in vivo can drive effective, specific downregulation of distinct target genes [11]. This feature is advantageous for functional analyses in vivo, for example, by addressing redundancy in gene families, or enabling combinatorial gene knockdowns to identify genetic interactions [25]. Mixing two cell type-specific plasmids before electroporation (*see* **Note 7**) can additionally be used to knock down many genes in distinct cell types, to rapidly investigate complex cellular and molecular interactions [25, 27]. The plasmids encoding the different miRNAs can be linked to the expression of fluorescent markers of different colors, which easily distinguishes the cells experiencing knockdown of the different target genes (Fig. 5).

Fig. 4 (continued) (βactin-mRFP1-miLuc), or miRNA against Contactin2 (β-actin-mRFP1-miCntn2). Transfected cells were identified by the expression of mRFP1 (red fluorescence, *upper row*), revealing similar transfection efficiencies in all conditions. Cells were then transfected with a construct driving the expression of chick Cntn2. Twenty-four hours later, the cells were fixed and immunolabeled for Cntn2. The relative expression level of Cntn2 (green fluorescence, *lower panels*) was used to assess knockdown efficiency of the miRNAs. See ref. [11] for details. (**b**) Testing miRNAs in vivo. In this example, chick neural tubes at HH17-18 were unilaterally electroporated with β-actin-hrGFPII-miLuc (*left panels*) or β-actin-hrGFPII-miCntn2 (*right panels*). Images show cross sections of the electroporated neural tubes at HH25. Dorsal is up. In situ hybridization for *Cntn2* was carried out. The electroporated hemi-segment of the spinal cord (identified by the expression of hrGFPII (green fluorescence, *insets*)) was compared to the non-treated half, which served as an internal control. While the electroporation of miLuc had no effect on the expression of *Cntn2*, *Cntn2* expression was clearly reduced in the presence of miCntn2 (*arrows*). The boxed areas are shown in higher magnification in the lower panels

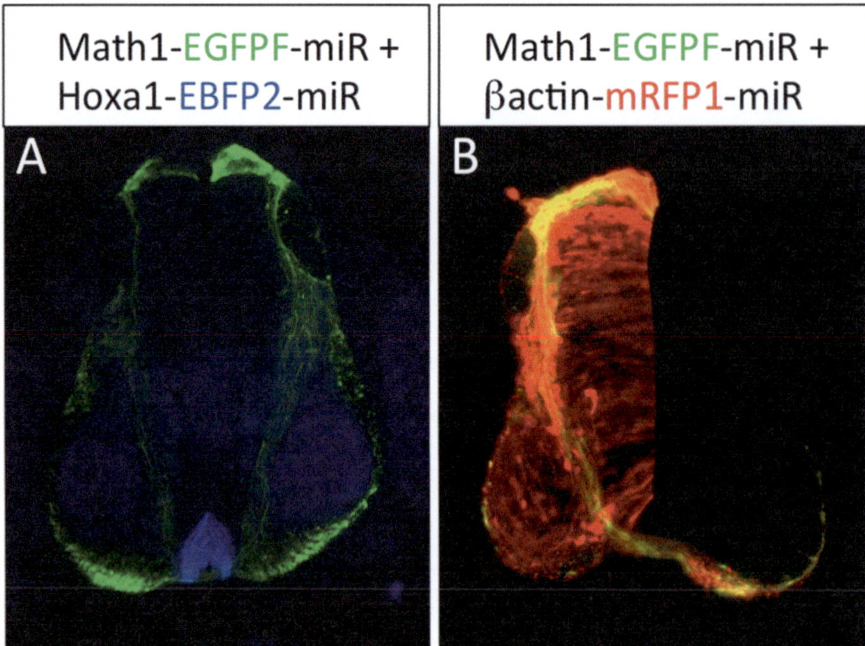

Fig. 5 Combinatorial application of miRNA constructs. Co-electroporation of miRNA constructs driven by different promoters/enhancers enables the simultaneous knockdown of multiple genes in distinct or overlapping cell types. Using a different fluorescent protein in each construct allows the easy identification of cells expressing each miRNA. (**a**) Embryo bilaterally co-electroporated with Math1-EGFPF-miR and Hoxa1-EBFP2-miR at HH17 displays *green* commissural axons and *blue* floor plate cells in spinal cord cross sections at HH25. (**b**) Embryo unilaterally co-electroporated with β-actin-mRFP1-miR and Math1-EGFPF-miR at HH17 displays widespread *red* cells and *green* commissural axons in spinal cord cross sections at HH25. Commissural axons expressing both constructs appear *yellow*

3.6 Phenotype Analyses

Depending on the scientific question, several techniques to analyze loss of function phenotypes are possible (Fig. 6). To monitor changes in cell migration or spinal cord patterning, immunohistochemistry and in situ hybridization are useful [4, 25, 28].

Defects in axon guidance can be detected by axonal tracing in slices [29, 30]. To visualize the peripheral nervous system and innervation of the limbs, whole-mount immunolabeling may be used [4, 24, 30]. We have also used a lipophilic fluorescent dye, DiI, to anterogradely label sensory neurons in the dorsal root ganglia [30, 31] and commissural neurons of the developing spinal cord [31]. The altered trajectory of DiI-labeled dI1 commissural axons is analyzed microscopically in "open-book" preparations of the spinal cord. We have previously described this dissection and mounting method in detail [17, 32]. In embryos electroporated with Math1-driven constructs, the miRNA-expressing commissural axon projections may be directly traced by the expression of the

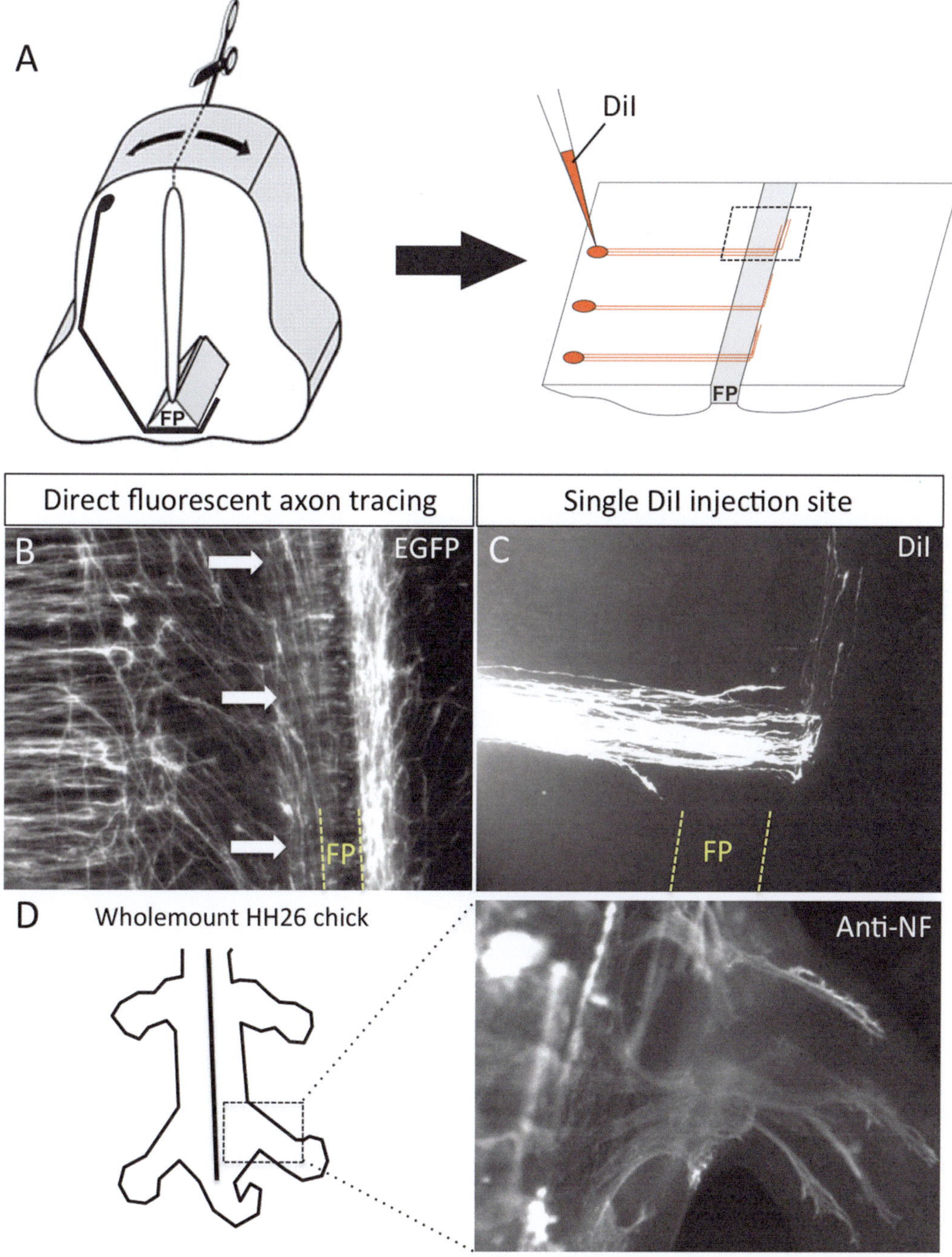

Fig. 6 Phenotype assessment techniques. (**a**) A schematic representation of the "open-book" preparation for analyzing commissural axon guidance defects. The spinal cord is dissected from the embryo, cut open along the roof plate, and opened into a flat-mount preparation. This allows the trajectory of the commissural axons to be visualized as they grow towards and across the floor plate (FP) at the ventral midline. (**b**) In embryos electroporated with Math1-driven miRNA constructs, the commissural axon trajectories may be directly traced by imaging the linked fluorescent reporter. In this example, the embryo was electroporated with

linked, membrane-bound fluorescent protein in sections or open-books [11].

The miRNA-expressing neurons may also exhibit changes in cell adhesion or outgrowth. These properties may be assessed in vitro by preparing cultures of dissociated neurons from electroporated embryos [27]. Because electroporation is not 100 % efficient [2], the pools of neurons obtained comprise both wild-type and miRNA-expressing (fluorescent) neurons from the same embryo. Thus, the wild-type neurons provide a convenient internal control in these assays.

4 Notes

1. The following backbone plasmids are available on request from the Stoeckli laboratory:
   (a) For ubiquitous expression:
   β-Actin promoter-mRFP1-miR
   β-Actin promoter-EBFP2-miR
   β-Actin promoter-hrGFPII-miR
   (b) For expression in dI1 commissural neurons:
   Math1-EGFPF-miR
   Math1-tdTomatoF-miR
   (c) For expression in the floor plate:
   Hoxa1-EGFP-miR
   Hoxa1-EBFP2-miR
2. The best test for specificity is the analysis of expression levels of non-targeted members of the same gene family. Appropriate negative controls (scrambled and/or unrelated sequences) are necessary. Rescue experiments should also be performed if possible [11, 20, 25, 27].
3. Many other online resources are available for artificial miRNA target prediction. This is a rapidly growing area of research and as such our knowledge of the factors affecting miRNA

Fig. 6 (continued) Math1-EGFPF-miCntn2 and imaged for EGFP using a 10× objective. While the axons grow correctly towards the floor plate (FP) at the midline, an abnormal fascicle of fluorescent axons grows along the ipsilateral floor plate border (*arrows*). (**c**) Alternatively, DiI injections into the cell bodies of the dI1 commissural neurons can be used to label smaller cohorts of axons at the midline. This example (40× objective) shows the normal phenotype in which dI1 commissural axons grow across the floor plate (FP) before turning and growing anteriorly. The region shown corresponds to the boxed area in (**a**). (**d**) To examine the peripheral nerves and limb innervation patterns, whole-mount immunolabeling may be used. In this example, anti-neurofilament-M (anti-NF; RMO270) reveals the nerves innervating the hindlimb of an HH25 embryo

efficiency and specificity is improving. This may be reflected in the different online algorithms available for miRNA design [33].

4. Dilute your 10× sequencing buffer into your sequencing reaction to make a 1× final concentration of 10 mM Tris–Cl with 0.01 mM EDTA (pH 8.0).
5. Elute the plasmid destined for in ovo electroporation in water rather than TE buffer, as TE buffer has been shown to cause electroporation artifacts [22]. Plasmids should be purified carefully, making sure to minimize alcohol contamination from previous precipitation or washing steps.
6. Use of a spectrophotometer can also help identify whether there are contaminating solvents or proteins in your DNA sample.
7. We recommend determining and using the lowest effective concentration of miRNA plasmids to suit the purpose. In our hands, the electroporation of high concentrations of plasmids with strong ubiquitous promoters (>1.0 μg/μl) is associated with increased incidence of nonspecific morphological abnormalities in the development of the neural tube. Exogenous miRNA and fluorescent protein may compete with the endogenous cellular machinery required for normal mRNA processing, localization and function. If the concentration of plasmids driven by cell-specific enhancers is too high, it may lead to "leakiness" of the enhancer activity and subsequent expression in non-target cells. Conversely, too little plasmid may reduce the expression of the linked fluorescent protein and prevent the electroporated cells from being traceable. As a guide, we use:
   (a) β-Actin promoter: 0.25 μg/μl
   (b) Math1 enhancer: 0.7 μg/μl
   (c) Hoxa1 enhancer: 0.5–1.0 μg/μl
8. Fertile eggs can be stored at 15 °C (in a climate controlled room, or, for smaller quantities of eggs, in a wine fridge) for up to 1 week to induce developmental stasis. Longer storage will impair the viability and development of the embryos. Eggs should be placed at this storage temperature as soon as possible after receiving them to maintain their quality. Leaving the eggs unattended at inappropriate ambient temperatures after their delivery (i.e., in cold winter or hot summer) will reduce embryo viability.
9. High humidity in the incubator (~45 %) is crucial. Place a tray of water containing 0.1 g/l of copper sulfate into the incubator. Copper sulfate prevents contamination of the water.
10. Many commercial egg incubators feature a semi-manual or automatic rotating function that has to be turned off, as

windowed eggs cannot be rotated. For incubation of 2–3 days (as described in this protocol), it is not necessary to rotate the eggs to promote their viability.

11. For efficient gene silencing, electroporations have to be performed prior to the onset of expression of your target gene in the cell type of interest.
12. Take precautions to reduce the risk of contamination in the egg. In addition to sterilizing your tools, solutions, and workspace, keep the time during which the egg is unsealed to a minimum.
13. Remove small batches of eggs from the incubator (6–10 eggs at a time) to perform the windowing and electroporations. This ensures that the temperature of the eggs in maintained within a narrow range.
14. Drilling a hole into the corner of the intended window not only provides a convenient insertion point for the scissors, but is also necessary to equilibrate the air pressure inside the egg and facilitate the detachment of the embryo from the eggshell during the removal of albumen.
15. If the yolk is damaged, the embryo is unlikely to survive. Eggs with damaged yolk should be discarded.
16. Make shallow cuts using only the tips of the scissors to avoid damaging the embryo or blood vessels. The window should be approximately 2×2 cm. If it is too small, access to the embryo will be limited.
17. After windowing, the eggs must be sealed properly. Loss of humidity in the egg is detrimental to the viability of the embryo.
18. Use a microscope to view the glass needle. Break off the tip at the end of the needle using forceps. If the tip diameter is too wide, it will be difficult to accurately insert the needle into the neural tube and your injection solution will spill out quickly, resulting in low and variable transfection efficiencies.
19. A commercially available egg slicer provides a convenient, washable, and reusable holder for the egg during electroporation. The slicing blades may be removed for safety. Alternatively, a standard egg carton may be cut to shape.
20. The membranes are "looser" around the middle of the embryo, in the region where the main blood vessels enter perpendicularly to the body axis. Use forceps to grasp and lift the membranes away from the embryo in this region. Use another pair of forceps or spring scissors to gently tear or cut the membranes towards the tail. It is not necessary to remove the membranes completely. A small clear area is sufficient. Removal of the vitelline membranes should not cause bleeding. Blood indicates

that the embryo or the surrounding blood vessels have been damaged, which should be avoided.

21. After HH18, the trunk of the chick embryo is rotated and no longer lies flat on the yolk [18]. For right-handed users, accessibility to the neural tube at these stages may be improved by positioning the embryo under the microscope so that the head is down and the tail is up. The red blood vessel running along the ventral side of the neural tube can be used to guide the insertion of your needle. The injection should be very shallow. If the needle is in the correct position, trypan blue dye should be seen extending in a thin line towards the tail and towards the head. Continue injecting until the entire neural tube is filled, but do not overfill. The heart rate of the embryo will slow if too much solution has been injected. Leave the needle in place for a few seconds before withdrawing it, to prevent spillage.
22. In our experience, mouth pipetting is the most efficient and controllable method for neural tube injections. However, if mouth pipetting is prohibited in the user's facility, a manual microinjection apparatus (e.g., Sutter Instrument) may be used.
23. With the embryo positioned in between the electrodes, press down gently until the electrodes are touching the yolk. Then lift the electrodes slightly to ensure that the underlying blood vessels are not in direct contact with the electrodes during electroporation.
24. To perform bilateral electroporations, use 18 V. After the first set of pulses, switch the polarity of the electrodes and repeat electroporation.
25. Successful electroporation will be evident by the formation of bubbles around the electrodes.
26. As an alternative to using Scotch tape, seal the window with a square cover slip (e.g., 22 $\times$ 22 mm) and paraffin wax. Apply melted paraffin to the edges of the window and carefully press a cover slip onto the wax until a proper seal is obtained. An advantage of this method is that the clear cover slip allows the embryo to be monitored on subsequent days without disturbing the seal. To remove the cover slip at the end of the experimental period, use a small soldering iron to melt the paraffin around the edges of the cover slip, and then carefully slide the cover slip off.
27. In a small number of cases, miRNAs that are successful in vitro are less effective in vivo [11, 34].
28. Embryos electroporated with cell type-specific constructs will be more difficult to assess by this pre-screening of whole embryos, owing to the smaller number of cells that will be

fluorescent. We recommend fixing and sectioning those embryos to verify electroporation efficiency and specificity by fluorescent protein expression.

29. As miRNAs may function at the level of transcript degradation and/or translational repression [35], an analysis method that detects protein levels of the target gene is best suited for determining miRNA knockdown efficiency. However, in the absence of antibodies against the gene of interest, mRNA detection techniques may be used.

References

1. Friedel RH, Wurst W, Wefers B, Kühn R (2011) Generating conditional knockout mice. Methods Mol Biol 693:205–231
2. Pekarik V, Bourikas D, Miglino N, Joset P, Preiswerk S, Stoeckli ET (2003) Screening for gene function in chicken embryo using RNAi and electroporation. Nat Biotechnol 21:93–96
3. Bourikas D, Pekarik V, Baeriswyl T, Grunditz A, Sadhu R, Nardó M, Stoeckli ET (2005) Sonic hedgehog guides commissural axons along the longitudinal axis of the spinal cord. Nat Neurosci 8:297–304
4. Mauti O, Domanitskaya E, Andermatt I, Sadhu R, Stoeckli ET (2007) Semaphorin6A acts as a gate keeper between the central and the peripheral nervous system. Neural Dev 2:28
5. Baeriswyl T, Stoeckli ET (2008) Axonin-1/TAG-1 is required for pathfinding of granule cell axons in the developing cerebellum. Neural Dev 3:7
6. Niederkofler V, Baeriswyl T, Ott R, Stoeckli ET (2010) Nectin-like molecules/SynCAMs are required for post-crossing commissural axon guidance. Development 137:427–435
7. Domanitskaya E, Wacker A, Mauti O, Baeriswyl T, Esteve P, Bovolenta P, Stoeckli ET (2010) Sonic hedgehog guides post-crossing commissural axons both directly and indirectly by regulating Wnt activity. J Neurosci 30:11167–11176
8. Katahira T, Nakamura H (2003) Gene silencing in chick embryos with a vector-based small interfering RNA system. Dev Growth Differ 45:361–367
9. Das RM, van Hateren NJ, Howell GR, Farrell ER, Bangs FK, Porteous VC, Manning EM, McGrew MJ, Ohyama K, Sacco MA et al (2006) A robust system for RNA interference in the chicken using a modified microRNA operon. Dev Biol 294:554–563
10. Kurreck J (2009) RNA interference: from basic research to therapeutic applications. Angew Chem Int Ed Engl 48:1378–1398
11. Wilson NH, Stoeckli ET (2011) Cell type specific, traceable gene silencing for functional gene analysis during vertebrate neural development. Nucleic Acids Res 39(20):e133. doi:10.1093/nar/gkr628
12. Lee S-K, Jurata LW, Funahashi J, Ruiz EC, Pfaff SL (2004) Analysis of embryonic motoneuron gene regulation: derepression of general activators function in concert with enhancer factors. Development 131:3295–3306
13. Uemura O, Okada Y, Ando H, Guedj M, Higashijima S-I, Shimazaki T, Chino N, Okano H, Okamoto H (2005) Comparative functional genomics revealed conservation and diversification of three enhancers of the isl1 gene for motor and sensory neuron-specific expression. Dev Biol 278:587–606
14. Avraham O, Hadas Y, Vald L, Zisman S, Schejter A, Visel A, Klar A (2009) Transcriptional control of axonal guidance and sorting in dorsal interneurons by the Lim-HD proteins Lhx9 and Lhx1. Neural Dev 4:21
15. Timmer J, Johnson J, Niswander L (2001) The use of in ovo electroporation for the rapid analysis of neural-specific murine enhancers. Genesis 29:123–132
16. Kieleczawa J (2006) Fundamentals of sequencing of difficult templates—an overview. J Biomol Tech 17:207–217
17. Wilson NH, Stoeckli ET (2012) In ovo electroporation of miRNA-based plasmids in the developing neural tube and assessment of phenotypes by DiI injection in open-book preparations. J Vis Exp 68:e4384. doi:10.3791/4384
18. Hamburger V, Hamilton HL (1992) A series of normal stages in the development of the chick embryo. 1951. Dev Dyn 195:231–272

19. Baeriswyl T, Mauti O, Stoeckli ET (2008) Temporal control of gene silencing by in ovo electroporation. Methods Mol Biol 442:231–244
20. Cullen BR (2006) Enhancing and confirming the specificity of RNAi experiments. Nat Methods 3:677–681
21. Krull CE (2004) A primer on using in ovo electroporation to analyze gene function. Dev Dyn 229:433–439
22. Croteau LP, Kania A (2011) Optimisation of in ovo electroporation of the chick neural tube. J Neurosci Methods 201(2):381–384. doi:10.1016/j.jneumeth.2011.08.012
23. Rao M, Baraban JH, Rajaii F, Sockanathan S (2004) In vivo comparative study of RNAi methodologies by in ovo electroporation in the chick embryo. Dev Dyn 231:592–600
24. Stepanek L, Stoker AW, Stoeckli E, Bixby JL (2005) Receptor tyrosine phosphatases guide vertebrate motor axons during development. J Neurosci 25:3813–3823
25. Wilson NH, Stoeckli ET (2013) Sonic hedgehog regulates its own receptor on postcrossing commissural axons in a glypican1-dependent manner. Neuron 79(3):478–491. doi:10.1016/j.neuron.2013.05.025
26. Sato F, Nakagawa T, Ito M, Kitagawa Y, Hattori M-A (2004) Application of RNA interference to chicken embryos using small interfering RNA. J Exp Zool A Comp Exp Biol 301:820–827
27. Andermatt I, Wilson NH, Bergmann T, Mauti O, Gesemann M, Sockanathan S, Stoeckli ET (2014) Semaphorin 6B acts as a receptor in post-crossing commissural axon guidance. Development 141(19):3709–3720. doi:10.1242/dev.112185
28. Chesnutt C, Niswander L (2004) Plasmid-based short-hairpin RNA interference in the chicken embryo. Genesis 39:73–78
29. Perrin FE, Rathjen FG, Stoeckli ET (2001) Distinct subpopulations of sensory afferents require F11 or axonin-1 for growth to their target layers within the spinal cord of the chick. Neuron 30(3):707–723
30. Frei JA, Andermatt I, Gesemann M, Stoeckli ET (2014) The SynCAM synaptic cell adhesion molecules are involved in sensory axon pathfinding by regulating axon-axon contacts. J Cell Sci 127:5288–5302. doi:10.1242/jcs.157032
31. Perrin FE, Stoeckli ET (2000) Use of lipophilic dyes in studies of axonal pathfinding in vivo. Microsc Res Tech 48:25–31
32. Wilson NH, Stoeckli ET (2014) Open-book preparations from chick embryos and DiI labeling of commissural axons. Bio-Protocol 4(13): e1176, http://www.bio-protocol.org/e1176
33. Witkos TM, Koscianska E, Krzyzosiak WJ (2011) Practical aspects of microRNA target prediction. Curr Mol Med 11(2):93–109. doi:10.2174/156652411794859250
34. Ehlert EM, Eggers R, Niclou SP, Verhaagen J (2010) Cellular toxicity following application of adeno-associated viral vector-mediated RNA interference in the nervous system. BMC Neurosci 11:20
35. Wilczynska A, Bushell M (2015) The complexity of miRNA-mediated repression. Cell Death Differ 22(1):22–33. doi:10.1038/cdd.2014.112

Neuromethods (2017) 128: 183–204
DOI 10.1007/7657_2016_11

Published online: 9 December 2016

Three Dimensional Cell Culture of Human Neural Stem Cells Using Polysaccharide-Based Hydrogels and Subsequent Bioanalyses

Geun-woo Jin, Weili Ma, and Won H. Suh

Abstract

Protocols for culturing neural stem cells (NSCs) are increasingly finding utilization for studying and growing of tissues that can appropriately model the neural regeneration processes. Two-dimensional (2D) plastic or glass surface-enabled mammalian cell cultures have been the platforms for performing in vitro cell cultures. Isolated mammalian cells, however, come from three-dimensional (3D) spaces, thus recapitulating such 3D microenvironments is among the challenges for many tissue engineering applications. Herein, we present the protocols for culturing NSCs in 3D polysaccharide-based hydrogel microenvironments that mimic, for instance, the native extracellular matrix (ECM) space (of the brain). The protocols include three key steps: (1) generation of 3D hydrogels with living cells, (2) culturing NSCs in 3D environments, and (3) characterization via immunostaining and genetic expression assay (RT-qPCR).

Keywords: Human neural stem cell, 3D cell culture, Calcium alginate, Hyaluronic acid, Quantitative Reverse transcription polymerase chain reaction (RT-qPCR), mRNA, miRNA

1 Introduction

In the field of neuroscience, one of the important milestones was the establishment of neural stem cells (NSCs), which are progenitor cells capable of generating neurons and glial cells. The discovery of NSCs shattered the dogma that the adult nervous system lacked any regenerative potential. Reynolds et al. isolated neural stem cells (NSCs) for the first time [1, 2] in 1992. They demonstrated that primary cells could be isolated from the central nervous system (CNS) of adult and embryonic mice and cultured in vitro. These NSCs were cultured in the presence of epidermal growth factor (EGF) and in the absence of extracellular matrix proteins to give rise to large spheres (few hundred microns) termed "neurospheres," which possessed a mixed population of neurons and glial cells. They demonstrated that a neurosphere could be derived from a single cell and could be dissociated back into single cells (suspended cells; like particles dispersed in liquid) to produce new neurospheres (and the process can be repeated). These neurosphere-producing cells exhibited (adult) stem cell properties that include self-renewal and multi-potent capabilities. In 1993, the studies of Luskin et al. [3] and Lois et al. [4] demonstrated that the proliferating cells in the adult rodent subventricular zone (SVZ) could differentiate into neurons within the olfactory bulb. Current theory holds that, within the adult SVZ, stem cells give rise to proliferating progenitors, which ultimately transform to neuronal precursors that migrate into the olfactory bulb to form granule cells and some periglomerular (PG) interneurons [5]. In addition to the olfactory bulb, the region in the adult hippocampus where new dentate gyrus cells are regularly generated is another known source of progenitor cells [6].

NSCs that persist in the adult mammalian brain can be cultured in vitro with growth factors as either adherent cultures or non-adherent cultures in the form of neurospheres [7, 8]. However, their biology is in large part defined by their in vivo niche (micro-environment). In the mammalian brain, there are two main NSC niches: the subventricular zone (SVZ) and the subgranular zone (SGZ) [9]. The SVZ is a paired brain structure situated throughout the lateral walls of the lateral ventricles. Newly generated neuroblasts traverse a network of chains which extends through the SVZ to join the rostral migratory stream (RMS), which then leads to the olfactory bulb. The neuroblasts differentiate into two kinds of interneurons, granular and periglomerular cells, and functionally integrate into the existing circuitry [10, 11]. The SGZ is located between the granule cell layer of the dentate gyrus and the hilus. Newly generated granule neurons in the SGZ are known to migrate to the granule cell layer, where they extend dendrites to the molecular layer and an axon along the mossy fiber path to integrate functionally into the circuitry of the dentate gyrus [12–14]. These native NSC niches are a dynamic and complex three dimensional (3D) microenvironment. Thus, the development of precisely defined NSC microenvironments is important for the increased understanding of how NSC biology can be controlled for in vitro studies. Recent biomaterials design strategies seek to create chemically well-defined 3D matrices that allow improved cell-cell interactions, cell metabolism, and chemical (i.e., oxygen, nutrition) transports occurring inside the 3D engineered space.

The most critical specific aim in all of 3D cell culture system development (for stem cells as well as other cell types) is assembling together defined molecular components that can most closely mimic (recapitulate) the native microenvironments (niches). From the viewpoint of the cells, it is basically enhancing the intercellular communications among stem cells and, in addition, achieving identical metabolic and nutritional regulation levels. The example molecular components, for such 3D systems, include materials derived from animals such as laminin and Matrigel™ that are currently available commercially for cell cultures [15–18].

While they are still widely utilized in various in vitro and in vivo experiments, they can incorporate residual and potentially detrimental components that are either not fully (chemically) defined or can affect animal biology (i.e., viral infection). This makes it difficult to conduct well-controlled studies with the bio-compounds and also pose problems when such materials are to be applied for actual human therapies. The most ideal 3D cell culture systems for cellular engineering work are those that are mainly fabricated utilizing completely well-identified (defined) chemicals (or materials). In recent years, polysaccharide-based hydrogels (i.e., hyaluronan, alginate) have been considered the prime base materials since they are chemically well-defined materials.

In this chapter, we want to introduce protocols for well-defined 3D stem cell culture systems along with their subsequent biological characterization details [19–22]. We will provide specific protocols developed in our lab. For instance, we are providing protocols that we utilized to minimize errors during PCR analyses for polysaccharide hydrogels, such as calcium alginate and radically crosslinked hyaluronan. It has been shown previously that polysaccharides can be PCR inhibitors depending on their incorporating chemical functionalities [23]. In order to address this issue, a newly modified method of nucleic acid isolation and purification using cetyltrimethylammonium bromide (CTAB) was developed [24, 25] specifically for human neural stem cells. A recent paper explored the current challenges in nucleic acid extraction from cells cultured within different types of hydrogels and compared various methods using commercially available assays, such as TRIzol and CTAB [26]. Although the specific 3D culture systems tested in this 2016 published article are different to the ones described in this chapter (i.e., barium alginate instead of calcium alginate), the overlying challenges with obtaining reproducible quantitative data from 3D cell cultures are apparent. We hope that this chapter gives the reader some insight into the current potentials and challenges regarding 3D cell culture systems for neural stem cell applications.

2 Materials

Calcein AM (CAT: 4892-010-01) was purchased from Trevigen. DAPI (4′,6′-Diamidino-2-phenylindole dihydrochloride, 98 %) was purchased from ACROS. ReNcell VM cells, ReNcell NSC maintenance medium, accutase, DMEM/F12, epidermal growth factor (EGF), basic fibroblast growth factor (bFGF), Stericup® filter (filter size: 0.22 μm), TRIzol, and antibodies were purchased from Millipore. Penicillin/streptomycin, laminin and paraformaldehyde (PFA), and 2-Hydroxy-4′-(2-hydroxyethoxy)-2-methylpropiophenone (Irgacure 2959 Photo-initiator) were purchased from Sigma Aldrich. Alginate powder and $CaCl_2$ were purchased from Willpowder. PureLink™ RNA mini kit was purchased from Ambion. Sodium hyaluronate (1.5 MDa) was purchased from Lifecore. Methacrylic Anhydride (94 % stab. w/ ca 0.2 % 2,4-dimethyl-6-tert-butylphenol) was purchased from Alfa Aesar. 3.5 kDa SnakeSkin dialysis tubing (22 mm) was purchased from Thermo Fisher Scientific. Sodium hydroxide was purchased from Fisher Chemical. Reagents for RT-qPCR including mRNA reverse transcription kit (Catalog number 4368814), miRNA reverse transcription kit (Catalog number A28007), TaqMan® Fast Universal Master Mix (Catalog number 4352042) for mRNA expression assay, TaqMan® Fast Advanced Master Mix (Catalog number 4444556) for miRNA expression assay, and gene-specific primer sets were purchased from Applied Biosystems.

3 Methods

3.1 Description of Immortalized NSCs

In our study, we used ReNcell VM [27], which is an immortalized human neural stem cell line with the ability to differentiate into neurons and glial cells. ReNcell VM was derived from the ventral mesencephalon region of a human fetal brain tissue. ReNcell VM was immortalized by retroviral transduction with the v-myc oncogene. This cell line is known to grow as a monolayer on laminin-coated surfaces and the doubling time is approximately 1 day (20–30 h). Even though karyotype analyses indicate that the ReNcell VM retains a normal diploid karyotype in culture even after prolonged passage (>45 passages), we selectively used ReNcell VM cells between passages 5 and 10.

3.2 Preparation of Laminin-Coated Flask

Laminin-coated tissue culture flasks (or glassware) are used to culture ReNcell VM cells. Tissue culture flasks should be coated before thawing ReNcell VM cells from liquid nitrogen. The laminin was thawed at 2–8 °C, then diluted with DMEM/F12 to 20 μg/mL. Enough laminin solution was added to cover the whole surface of the tissue culture flasks (3 mL for T25 flask and 6.5 mL for T75 flasks). Then, the flasks were incubated in the incubator (37 °C, 5 % CO_2) for 4 h. Before seeding the cells, the laminin solution was aspirated and the laminin-coated flask was rinsed once with PBS.

3.3 Human Neural Stem Cell Seeding and Proliferation

The cell culture medium and laminin-coated tissue culture flask should be prepared before thawing the cells. The complete ReNcell VM maintenance medium was prepared by adding 20 ng/mL bFGF, 20 ng/mL EGF, and 1 % penicillin/streptomycin to ReNcell NSC maintenance medium. We added 10 mL of the freshly prepared complete ReNcell VM medium to the laminin-coated T75 flasks (5 mL to T25 flasks).

After preparation of the flask filled with medium, we removed the vial of ReNcell VM cells from liquid nitrogen and incubated in a 37 °C water bath to thaw the cells (Closely monitor until the cells are completely thawed). As soon as the cells were completely thawed, they were transferred to a sterile 15 mL conical tube in a laminar flow hood. Then, we slowly added 9 mL of ReNcell NSC maintenance medium (pre-warmed to 37 °C) to the 15 mL conical tube. We mixed the cell suspension by slow pipetting up and down twice (Do not introduce any bubbles and do not vortex the cells). Then, we centrifuged the tube at 500 × *g* for 5 min to pellet the cells. The supernatant was carefully aspirated to remove residual cryopreservative (DMSO). The cells were resuspended in 5 mL of complete ReNcell VM maintenance medium (pre-warmed to 37 °C) and transferred to the laminin-coated T75 flask filled with 10 mL of complete ReNcell VM maintenance medium, so that the final volume was 15 mL. The cells were incubated at 37 °C in a 5 % CO_2 humidified incubator. On the next day, the medium was completely refreshed with complete ReNcell VM maintenance medium (pre-warmed to 37 °C). The medium was changed with fresh complete ReNcell VM maintenance medium every day thereafter. When the cells were approximately 80–90 % confluent, they were dissociated with Accutase and passaged (or alternatively frozen for later use).

3.4 Neural Stem Cell Seeding and Differentiation in 3D Hydrogels

3.4.1 Calcium Alginate System

Before the cell seeding in 3D alginate gels, sterile alginate solution (2 % w/v) was prepared by filtration with Stericup® filters (pore size: 0.22 μm). For the cell seeding in alginate hydrogels, one million cells were dispersed in 0.5 mL of complete ReNcell VM maintenance medium. The cell solutions were mixed with equal volume of 2 % w/v alginate solution to make a 1 % w/v alginate-cell mixture (final concentration one million of cells per mL). The alginate-cell solution was added drop-wise to a $CaCl_2$ solution (50 mM) inside 6-well plates to form spherical alginate beads (Fig. 1). The seeded cells were stabilized for 5 days in a complete ReNcell maintenance medium. The medium was replaced with fresh complete ReNcell maintenance medium every day. Then, the medium was changed to ReNcell VM differentiation medium (complete ReNcell VM maintenance medium without growth factors) to induce the differentiation of the stem cells. During the differentiation, the medium was replaced every other day with differentiation medium.

3.4.2 Methacrylated Hyaluronic Acid System

The other type of hydrogel tested in our lab is a hyaluronic acid (HA) gel, which was prepared by the chemical modification of HA with methacrylic anhydride (MA) and subsequent cross-linking [28, 29]. To prepare the methacrylated hyaluronic acid (MeHA), sodium hyaluronate (1.5 MDa, Lifecore) was dissolved in nanopure water (1 % w/v) by stirring at room temperature. The solution was

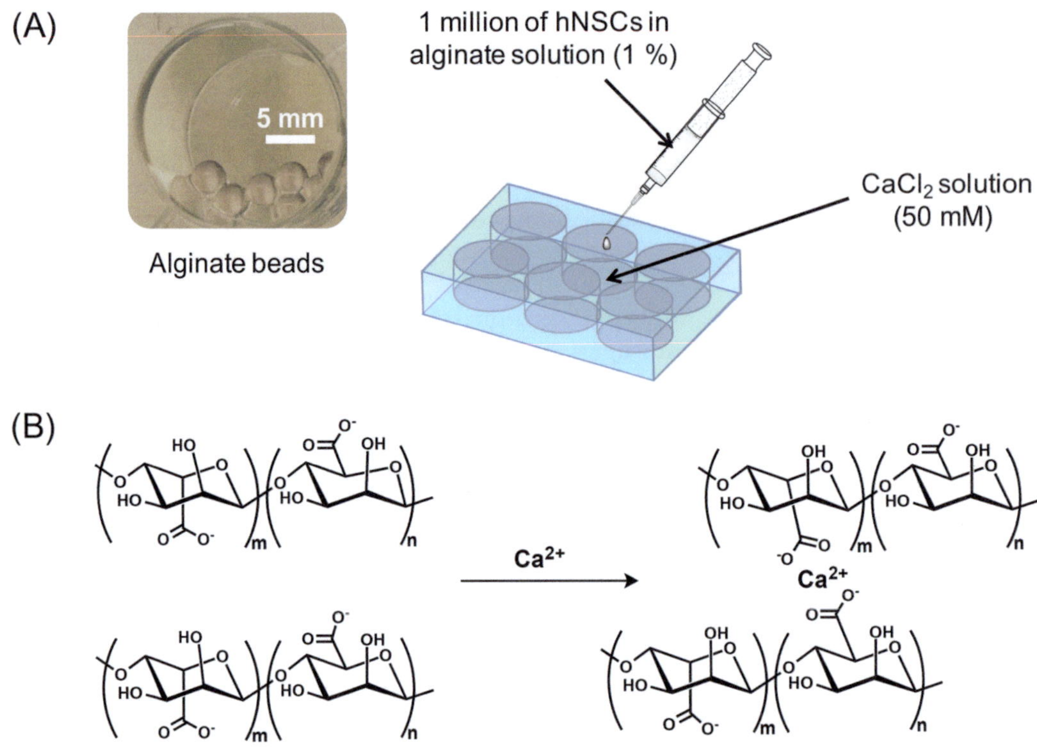

Fig. 1 (**a**) Procedure overview and gel photo, (**b**) Chemical reaction showing the formation (cross-linking) of calcium alginate

chilled on ice and adjusted to pH > 8 with 5 N NaOH. MA was added drop-wise at a 20 M excess ratio to the primary C_6 hydroxyl on the *N*-acetylglucosamine subunits of HA. During this process, the pH was monitored and adjusted as necessary with 5 N NaOH. The MeHA solution was protected from light with aluminum foil and transferred to 4 °C to continue stirring for a total of 24 h. The methacrylated HA (MeHA) solution was transferred to 3.5 kDa molecular weight cutoff dialysis tubing and dialyzed against nanopure water for 3 days, changing the water daily. The dry product was obtained by lyophilization. The MeHA was reconstituted in phosphate buffered saline (1 % w/v) by stirring at room temperature. Irgacure 2959 photo-initiator was added to the MeHA solution (0.05 % w/v) and was dissolved in the dark by covering the sample with aluminum foil. The Irgacure 2959 can be activated to initiate radical polymerization by exposure to 360 nm UV light [28]. The precursor solutions were transferred into 48-well molds for cell experiments. Cross-linked hydrogels were formed following exposure to a metal halide lamp (EXFO X-Cite 120Q) for 10 s. Due to the cytotoxicity of the photo-initiator and the photo-toxicity during exposure to a high powered light source, the hydrogels were made into porous sponge scaffolds by lyophilization prior to cell seeding. MeHA hydrogels were prepared in 48-well plates and frozen at −80 °C. Porous, cross-linked HA disks were generated by lyophilization. The dried hydrogel disks were taken out of the 48-well plates and placed into 24-well plates for cell experiments. Sterilization was performed by UV exposure for 30 min (15 min each side). 2 × 10^5 ReNcell VM cells were seeded onto each of the lyophilized HA disks. After the addition of complete ReNcell VM maintenance medium to a total volume of 200 μL, the cells were placed into the incubator at 37 °C (5 % CO_2 humidified) for attachment. After 1 h, the cell-laden hydrogel was swollen with medium and so 200 μL of additional complete ReNcell VM maintenance medium containing growth factors was added (Fig. 2). To initiate differentiation by growth factor withdrawal, excess growth medium was first removed from the wells by pipetting. The hydrogels were then rinsed and replaced with 400 μL ReNcell VM differentiation medium and placed back into the incubator

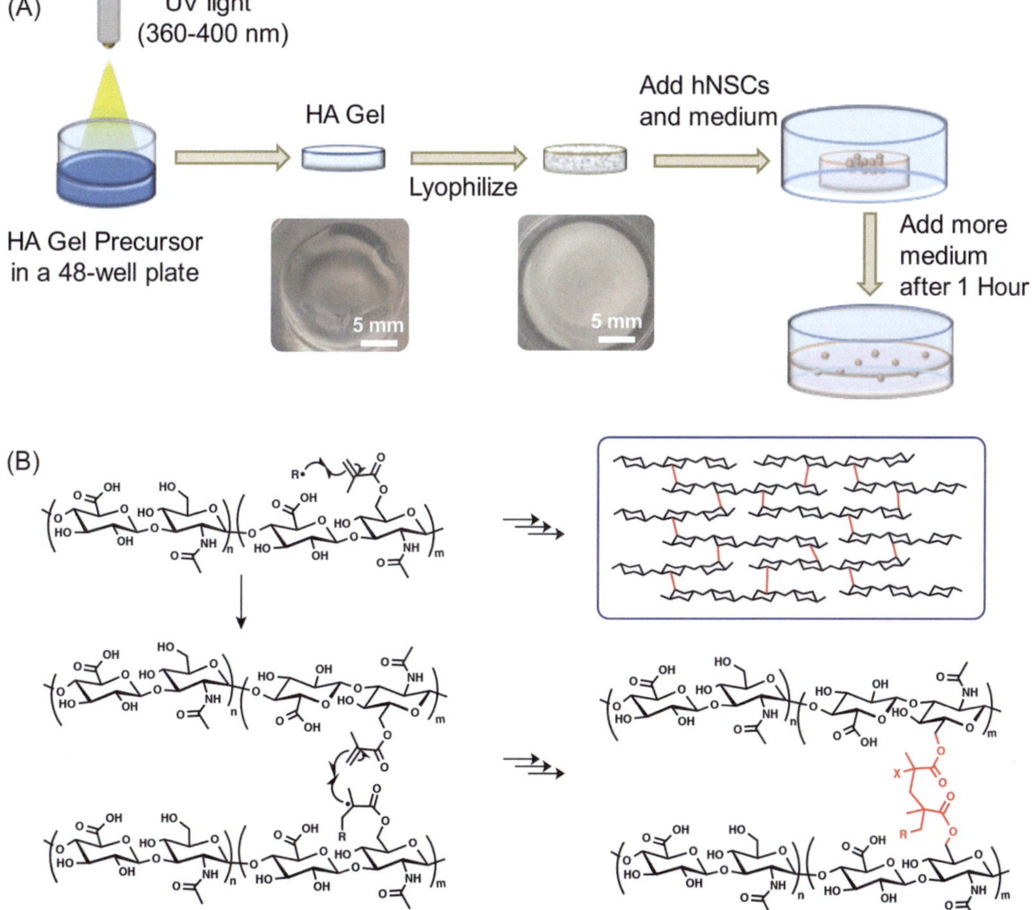

Fig. 2 (**a**) Overall scheme, (**b**) Main cross-linking chemistry of methacrylated hyaluronan synthesis

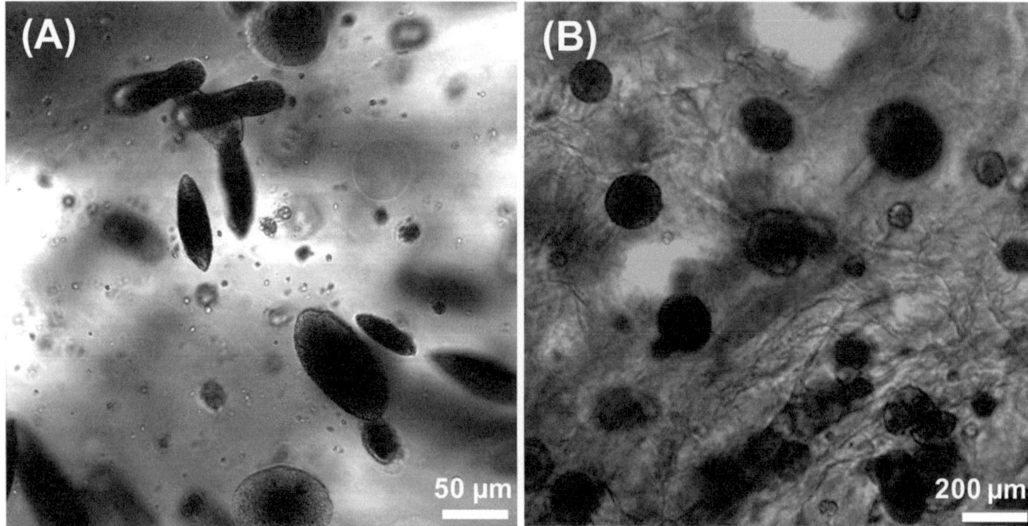

Fig. 3 NSCs seeded in (**a**) alginate (10× image) and (**b**) hyaluronan hydrogel (4× image)

(37 °C, 5 % CO_2). During the differentiation process, the medium was changed every day.

After seeding NSCs in the calcium alginate or MeHA hydrogels, the cells were observed with a confocal microscope fitted with a live-cell imaging device every day to monitor the cell growth. As shown in Fig. 3, hNSCs encapsulated in 3D gels proliferated as neurospheres.

3.5 Neural Stem Cell Isolation from 3D Alginate Gels

For the collection of cell lysate, cells were isolated from 3D alginate gels. Alginate beads were dissolved by adding 2 mL of sodium citrate solution (100 mM). The mixture (total volume: 3 mL) was mixed with 1 mL of complete ReNcell maintenance medium and transferred to a laminin-coated 6-well plate. Incubation for 1 h at 37 °C allowed cell attachment to the laminin-coated surface. The attached cells were rinsed with 4 mL of sodium citrate solution (two times) and PBS (two times) (Fig. 4). For RT-qPCR, the cells were lysed by the treatment of 1 mL TRIzol or CTAB buffer (100 mM Tris, 20 mM EDTA, 140 mM NaCl, 2 % CTAB, 1 % polyvinyl pyrrolidone 40,000) per well for 5 min.

3.6 Collection of Cell Lysate from NSCs Embedded in MeHA

On the second to last day of differentiation, 1000 U/mL bovine hyaluronidase was added to the differentiation medium to enzymatically digest the MeHA hydrogels. The hydrogels were completely dissolved after overnight treatment. The cell suspension was transferred to a sterile 1.5 mL centrifuge tube and pelletized by centrifuging at 500 × g for 5 min. The supernatant was removed and the cells were lysed for total RNA extraction.

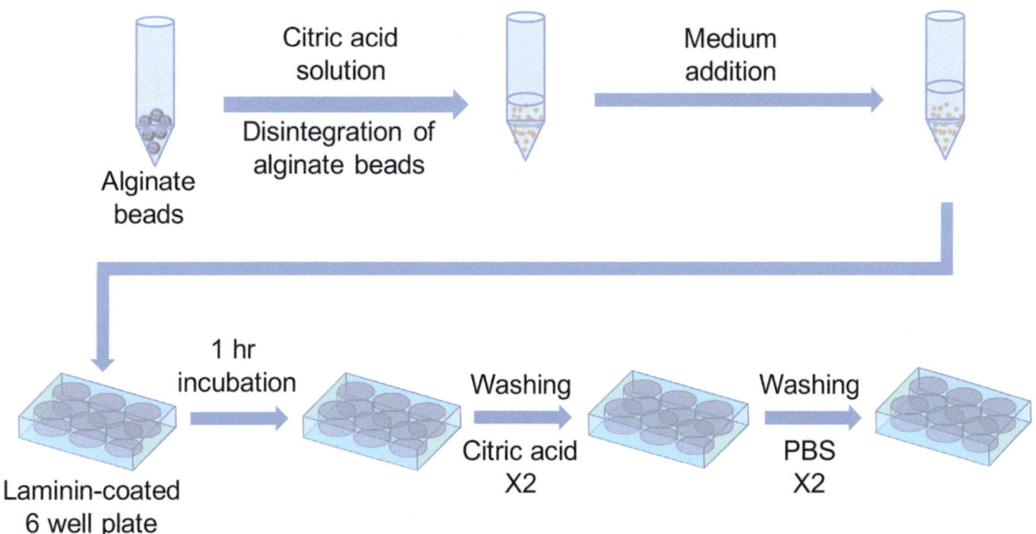

Fig. 4 The isolation of human NSCs by dissolving alginate beads

3.7 Total RNA Collection

Total RNA was collected from cell lysate using a PureLink™ RNA mini kit according to the manufacturer's protocol (https://tools.thermofisher.com/content/sfs/manuals/purelink_rna_mini_kit_man.pdf). Briefly, we added 0.2 mL chloroform to 1 mL TRIzol reagent or CTAB buffer used to lyse ReNcell VM cells. We capped and shaked the tube vigorously by hand for 15 s (Avoid vortexing to prevent DNA contamination of the samples). Then, the mixture was incubated at room temperature for 3 min. Then, we centrifuged the sample at 12,000 × *g* for 15 min at 4 °C. The colorless, upper phase was transferred to a new RNase-free tube (After centrifugation, the mixture separates into a colorless upper aqueous phase containing RNA, an interphase, and a lower phenol-chloroform phase). Equal volume of 100 % ethanol was added to the aqueous phase extract. The tube was vortexed and inverted to disperse RNA precipitates. The mixture was (≤700 μL) transferred to a spin cartridge (with a collection tube) from the kit. We centrifuged the tube at 12,000 × *g* for 1 min at room temperature and discarded the flow-through (If there is any remaining mixture, transfer to a Spin cartridge and repeat the centrifugation and discard the flow-through). The Spin cartridge was inserted into a new collection tube. We added 500 μL Wash buffer II (provided in the kit) with ethanol to the spin cartridge. The tube was centrifuged at 12,000 × *g* for 30 s at room temperature and we discarded the flow-through. We repeated the washing step with Wash buffer II with ethanol. After centrifugation at 12,000 × *g* for 1 min at room temperature, we dried the membrane with bound RNA for 1 min. We discarded the flow-through and the collection tube and inserted the spin cartridge into a recovery tube. We added 100 μL of RNase-free water to the center of the spin cartridge and incubated at room temperature for 1 min. We centrifuged the spin cartridge with the recovery tube for 2 min at 12,000 × *g* at room temperature. The concentrations of mRNA samples were evaluated based on optical density at 260 nm (OD_{260}) and the purity was determined by OD_{260}/OD_{280}. The mRNA samples having OD_{260}/OD_{280} values of 1.8–2.1 were considered pure for conducting RT-qPCR.

3.8 Preparation of cDNA from mRNA/miRNA

3.8.1 Reverse Transcription of mRNA

The RNA samples and reagents were thawed on ice. The reaction cocktail of the following reagents was prepared in a microcentrifuge tube (Table 1).

The reaction cocktail was added to the 10 μL RNA sample and vortexed to mix the reagents. Any air bubbles were eliminated with centrifugation (200 × *g*, 1 min). After sealing the plate/tube(s), they were place in a thermal cycler and incubated with the following settings (Table 2).

3.8.2 Reverse Transcription of Noncoding RNA (miRNA)

Micro-RNAs (miRNAs) are a group of 20- to 24-nucleotide small RNAs that play an important role in regulating genetic expression. However, the small size of miRNAs makes them difficult to be detected and analyzed by conventional RT-qPCR techniques. So, additional steps are required for the reverse transcription of miRNA. We followed the miRNA reverse transcription process as recommended by Thermo Fisher Scientific (https://tools.thermofisher.com/content/sfs/manuals/100027897_TaqManAdv_miRNA_Assays_UG.pdf). In this method, total RNAs, including miRNAs, are extended by a poly(A) tailing reaction using poly(A)

Table 1
Reaction cocktail for reverse transcription of mRNA

Component	Volume (μL)
10× RT buffer	2.0
25× dNTP mix (100 mM)	0.8
10× RT random primers	2.0
MultiScribe™ Reverse transcriptase	1.0
Nuclease-free water	4.2
Total volume	10.0

Table 2
Thermal profile for reverse transcription of mRNA

	Step 1	Step 2	Step 3	Step 4
Temperature (°C)	25	37	85	4
Time	10 min	120 min	5 min	∞

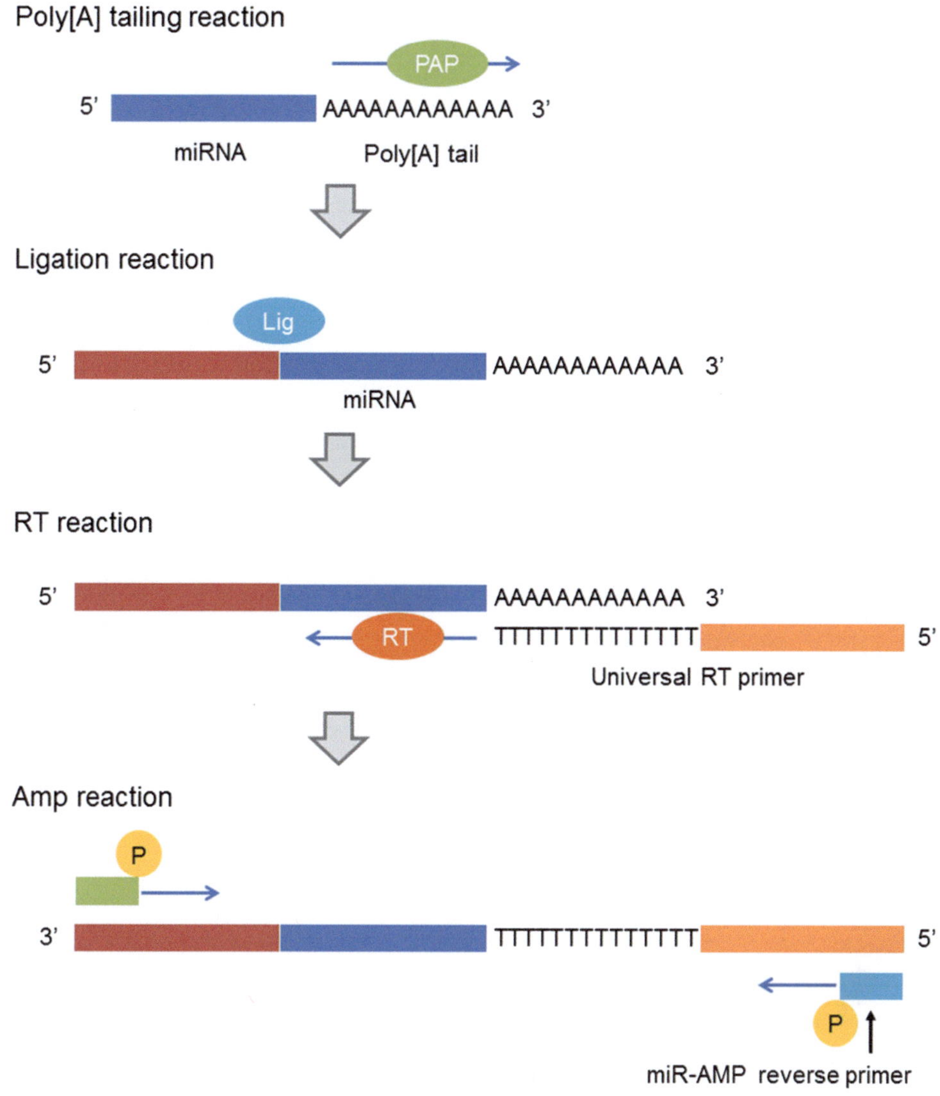

Fig. 5 Process for reverse transcription of miRNA. *PAP* Poly[A] enzyme (polymerase), *Lig* RNA ligase, *RT* reverse transcriptase

polymerase and ligation. The extended miRNA is converted into cDNA through reverse transcription primed by a poly(T) adaptor, and then PCR-amplified to increase the number of cDNA (Fig. 5).

3.9 Poly[A] Tailing

We added 2 μL of total RNA sample to each well of a reaction plate and mixed with 3 μL of the poly[A] tailing reaction cocktail (Table 3). Then, the plate was sealed and vortexed briefly (3 s). The sealed plate was centrifuged to spin down the contents and eliminate any air bubbles. The plate was placed into a thermal cycler and incubated with the following settings (Table 4).

Table 3
Reaction cocktail for poly[A] tailing

Component	Volume (μL)
10× poly[A] buffer	0.5
ATP	0.5
Poly A enzyme	0.3
Nuclease-free water	1.7
Total volume	3.0

Table 4
Thermal profile for poly[A] tailing

	Step 1	Step 2	Step 3
Temperature (°C)	37	65	4
Time	45 min	10 min	∞

Table 5
Reaction cocktail for ligation

Component	Volume (μL)
5× DNA ligase buffer	3.0
50 % PEG 8000	4.5
25× ligation adaptor	0.6
RNA ligase	1.5
Nuclease-free water	0.4
Total volume	10.0

3.10 Ligation

After poly[A] tailing, 10 μL of the ligation reaction cocktail (Table 5) was added to each well of a reaction plate containing the poly(A) tailing reaction product (15 μL total volume in each well/tube) then sealed and vortexed (3 s). The plate was centrifuged briefly to pull down reagents. The plate was placed into a thermal cycler and incubated with the following setting (Table 6).

3.11 RT Reaction

RT reaction cocktail (Table 7) was added to each well of a reaction plate containing the ligation reaction product (30 μL total volume in each well/tube). Then, the plate was sealed and the reagents

Table 6
Thermal profile for ligation

	Step 1	Step 2
Temperature (°C)	16	4
Time	60 min	∞

Table 7
Reaction cocktail for RT reaction

Component	Volume (μL)
5× RT buffer	6.0
dNTP mix	1.2
20× Universal RT primer	1.5
10× RT enzyme mix	3.0
Nuclease-free water	3.3
Total volume	15.0

Table 8
Thermal profile for RT reaction

	Step 1	Step 2	Step 3
Temperature (°C)	42	85	4
Time	15 min	5 min	∞

were mixed thoroughly by vortexing (3 s). The plate was centrifuged to spin down the contents and placed into a thermal cycler with the following settings (Table 8).

3.12 Amp Reaction

We added 5 μL of the RT reaction product to each well of the plate containing 40 μL Amp reaction cocktail (Table 9). The plate was sealed and vortexed briefly (3 s). Then, the plate was centrifuged to spin down the contents and placed into a thermal cycler with the following settings (Table 10).

3.13 RT-qPCR

PCR amplification of cDNA was carried out using a universal PCR master mix (Applied Biosystems) and gene-specific primer sets (Thermofisher Scientific). For mRNA expression assay, 2 μL of cDNA prepared from mRNA was mixed with reaction cocktail

Table 9
Reaction cocktail for Amp reaction

Component	Volume (μL)
2× miR-Amp master mix	25
20× miR-Amp primer mix	2.5
Nuclease-free water	17.5
Total volume	45

Table 10
Thermal profile for Amp reaction

	Step 1	Step 2	Step 3	Step 4	Step 5
Temperature (°C)	95	95	60	99	4
Time	5 min	3 s	30 s	10 min	∞

Table 11
Reaction cocktail for mRNA expression assay

Component	Volume (μL)
TaqMan® Fast Universal Master Mix	5.0
TaqMan® assay mix (primer)	0.5
Nuclease-free water	2.5
Total volume	8.0

(Table 11). Reaction cocktail for miRNA expression assay was prepared as described in Table 12 and mixed with 2.5 μL of cDNA prepared from miRNA. The genetic expression profile for mRNA or miRNA was generated using the thermal cycler with the following settings (Table 13).

For mRNA expression assay, we used primers for neural stem cell (Nestin, SOX2), immature neuron (Tubb3), mature neuron (MAP2), and astrocyte (GFAP) specific genes. GAPDH was used as a control (house-keeping gene) (Fig. 6). Primers for miRNA expression assay include miR-191-5p, miR-423-5p, and miR-124-3p. miR-124 has been known to regulate neuronal differentiation, while miR-191 and miR-423 were internal standards (Fig. 7). The cycles for RT-qPCR consisted of 1 s denaturation at 95 °C and 20 s annealing at 60 °C. RT-qPCR products were

Table 12
Reaction cocktail for miRNA expression assay

Component	Volume (μL)
TaqMan® Fast Advanced Master Mix	5.0
TaqMan® assay mix (primer)	0.5
Nuclease-free water	2.0
Total volume	7.5

Table 13
Thermal cycle profile for RT-qPCR

	Enzyme activation	40 cycles	
Temperature (°C)	95	95	60
Time (s)	20	1	20

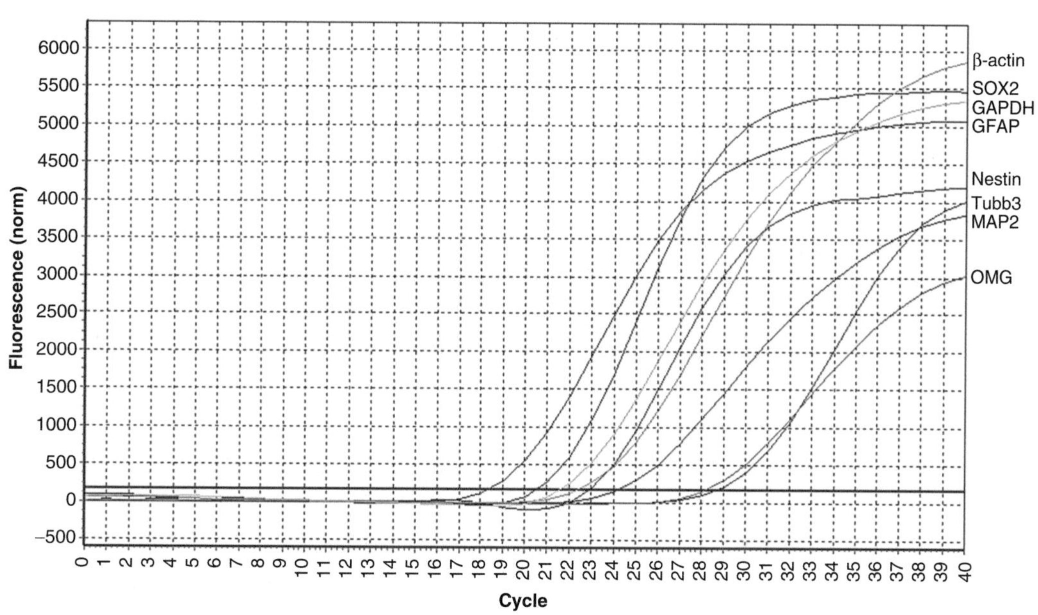

Fig. 6 Example RT-qPCR amplification plots of Nestin, SOX2, Tubb3, MAP2, GFAP, OMG, β-actin, and GAPDH genes from hNSCs cultured in 3D alginate gels (5 days of proliferation in the presence of growth factors and 14 days of differentiation in growth factor depleted condition)

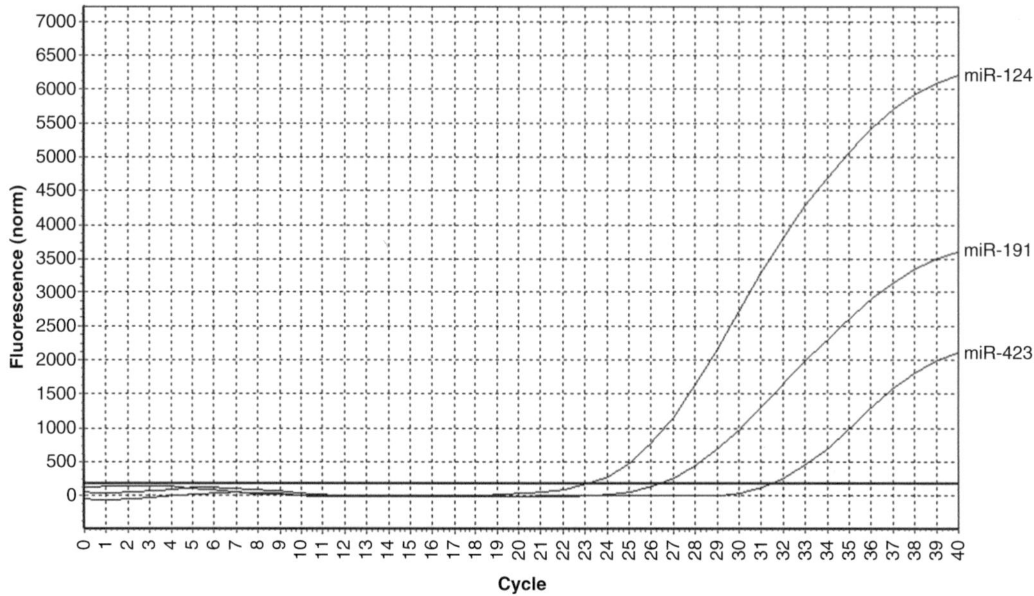

Fig. 7 Example RT-qPCR amplification plots of miR-124, miR-191, and miR-423 from hNSCs cultured in 3D alginate gels (5 days of proliferation and 7 days of differentiation, 0.1 μM of RA was treated)

analyzed using Mastercycler ep realplex 4 (Eppendorf). Three repeats of the experiments should be conducted to perform statistical analysis such as *t*-tests and ANOVA.

3.14 Immunostaining (ICC)

Cells were fixed for immunocytochemistry staining with corresponding markers. The medium was removed from the wells and the cell-laden hydrogels were rinsed with PBS three times. On the last wash, the PBS was left in the hydrogels for 30 min. Afterward, the hydrogels lost their color due to the medium and became transparent. The cell-laden hydrogels were incubated at room temperature with 4 % paraformaldehyde for 15 min. The samples were then rinsed with warm PBS using the same method as above. Blocking and permeabilization solution was prepared by diluting donkey serum (3 % v/v) and Triton-X 100 (0.3 % v/v) in PBS. The blocking and permeabilization solution was filtered through a 0.22 μm membrane prior to use. The samples were blocked and permeabilized for 30 min and then rinsed with PBS using the above method. Primary antibodies were diluted (1:200) in blocking solution (3 % v/v donkey serum in PBS, filtered) and added to the samples. The primary antibodies were incubated for 2 h at room temperature and then the cells were rinsed with PBS using the above method. Secondary antibodies were added using the same method as primary antibodies but with the samples covered in foil to prevent photo-bleaching. The samples were rinsed a final time with PBS using the above method and kept hydrated in PBS at 4 °C until microscopy analysis.

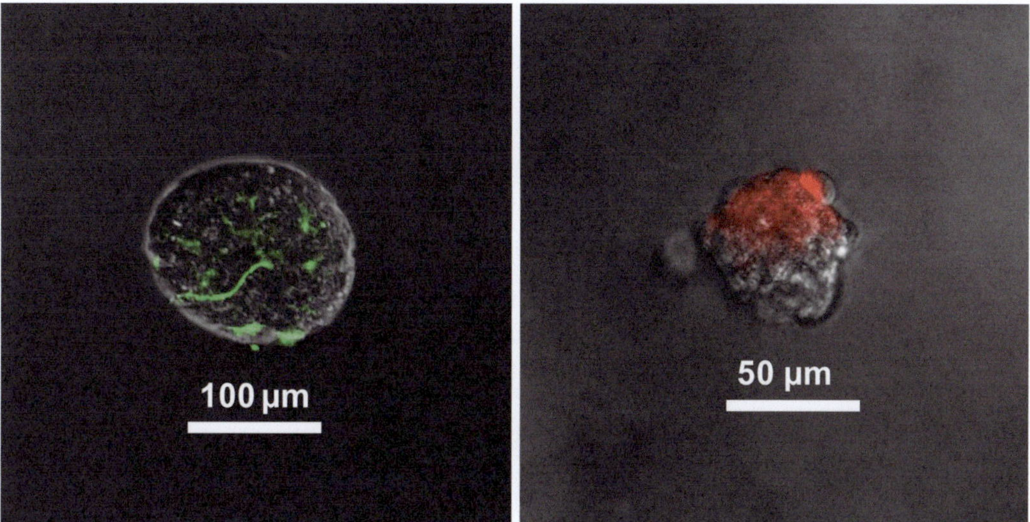

Fig. 8 Immunostaining images of hNSCs in 3D alginate (*left*) and HA (*right*) gels after 7 days of differentiation. Neurons derived from hNSCs were stained with anti Tubb3-Alexa Fluor® 488 (*left*, dye conjugated primary antibody) and anti Tubb3/Alexa Fluor® 555 (*right*, primary and secondary, respectively)

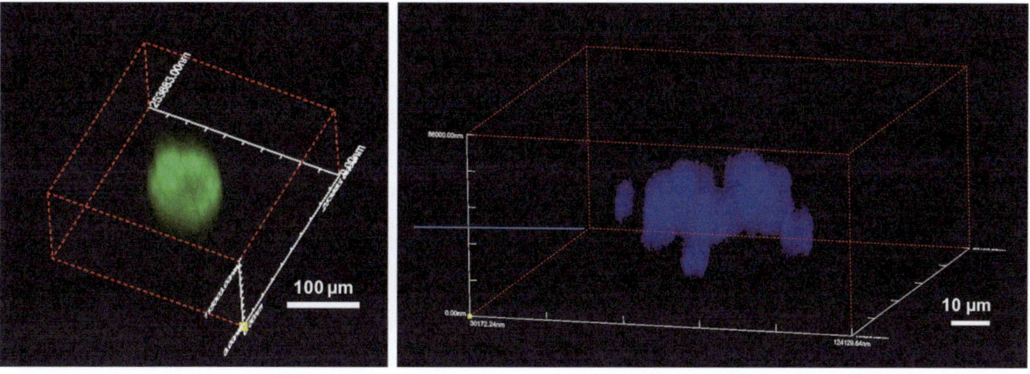

Fig. 9 3D reconstruction acquired from Z-stacking. NSCs in alginate (*left*) and HA (*right*) gels were stained with calcein AM (2 μM) and DAPI, respectively

3.15 Microscopy

Cells encapsulated in gels were imaged using an inverted bright-field microscope (Olympus CKX-41) (Fig. 3). Fluorescence was imaged (Fig. 8) using a wide-field spinning disc confocal fluorescent microscope (Olympus IX-83 DSU) or a laser scanning confocal microscope (Olympus FV1200). Z-stacks were obtained by taking few hundred μm slices within the hydrogels at 1–2 μm intervals. 3D reconstruction was performed either using the Olympus CellSens software or the FV1200 LSCM software after imaging on the microscope (Fig. 9).

4 Notes

This book chapter describes the establishment of 3D polysaccharide-based hydrogels as in vitro culture systems for human NSCs. The polysaccharides have been considered ideal matrices for in vitro culture models due to their biocompatibility, mechanical properties, and defined chemical constituents. However, there are some important points to consider when culturing cells in a three-dimensional space.

First, the continuous cell growth in hydrogel is dependent on the diffusion, transport, and utilization of nutrients and oxygen by the cells. If such molecular components do not reach cells inside the 3D gel space in a timely manner, cells may experience necrosis and eventual cell death. The NSCs grown in the gel forming neurospheres (Fig. 8) imply that the polysaccharide-based hydrogels permitted free diffusion of nutrients and growth factors inside the 3D hydrogel. Second, the toxic potentials of molecular and ionic components produced during the 3D hydrogel synthesis process will need to be carefully evaluated. Due to the considerable cytotoxicity of the photo-initiator and the photo-toxicity during exposure to a high powered light source (e.g., 120 mW), modified HA was cross-linked and then porous scaffolds were prepared by lyophilization prior to cell seeding. Although this photoinitiator system has been known to be compatible with other cell types [30], it has a very low viability with neural progenitor cells [31].

For alginate gel formation, we cross-linked alginate with calcium ion (e.g., 10–100 mM), which is known to damage the cell membrane [32]. To reduce cell death, the calcium chloride concentration was adjusted to 50 mM and the treatment time was limited to 10 min. After encapsulating NSCs, excess calcium chloride was rinsed off by the medium. Third, when working in a 3D network, cells have limited accessibility, also, and isolation of genetic materials from the gels can be difficult. The PCR inhibition was observed for human NSCs isolated from both alginate and HA gels. Previous studies in plant biology and analytical chemistry have shown that polysaccharides in plant tissues inhibit the activity of *Taq* polymerase by interfering with the binding of Mg^{2+} [33–35]. Westhrin et al. proposed stem cell isolation method with centrifugation after dissolving alginate gels to solve this problem [36]. They added citrate solution to the alginate gel and vortexed (5 min) to dissolve the gel. Then, they centrifuged (3 min, 1000 rpm) to recover the mesenchymal stem cells (MSCs). Subsequently, the recovered cells were washed with PBS to remove alginate from the sample. However, we found that this centrifugation method was not applicable in isolating hNSCs because of the low cell isolation efficiency. Most cells (>80 %) were lost after first round of centrifugation and no cells were found after the second round.

Herein, we proposed a cell isolation protocol for hNSCs to remove polysaccharide (alginate) while minimizing cell loss through cell reseeding on laminin-coated plates and washing with citric acid solution as shown in Fig. 3. However, we could only get complete Ct value data set for gene markers such as Nestin, SOX2, Tubb3, MAP2, GFAP, OMG, β-actin, and GAPDH with a success rate of 25 %. This result is not due to the low RNA isolation efficiency since we could collect RNA over 30 ng/μL with acceptable purity ($OD_{260}/OD_{280} > 1.8$) from hNSCs in 3D alginate gel (Table 14). This potentially is due to the remnant alginate

Table 14
Concentration and purity of RNA extracted from hNSCs in 3D alginate gels by TRIzol (top) or CTAB method (bottom)

	Concentration (ng/μL)	OD_{260}/OD_{280}
Sample 1	53.20	1.97
Sample 2	42.00	1.92
Sample 3	52.32	1.81
Sample 4	38.08	1.93
Sample 5	72.32	2.07
Sample 6	46.32	1.98
Sample 7	35.92	1.98
Sample 8	41.44	1.89
Sample 9	61.92	2.00
Sample 10	46.72	2.02
Sample 11	51.84	2.01
Sample 12	69.36	2.06
	Concentration (ng/μL)	OD_{260}/OD_{280}
Sample 1	35.44	1.92
Sample 2	31.84	1.98
Sample 3	34.16	1.89
Sample 4	60.40	1.99
Sample 5	64.64	1.97
Sample 6	48.64	2.00
Sample 7	54.64	1.90
Sample 8	32.08	1.97
Sample 9	45.12	1.91
Sample 10	41.76	1.98
Sample 11	38.72	1.96
Sample 12	36.00	1.97

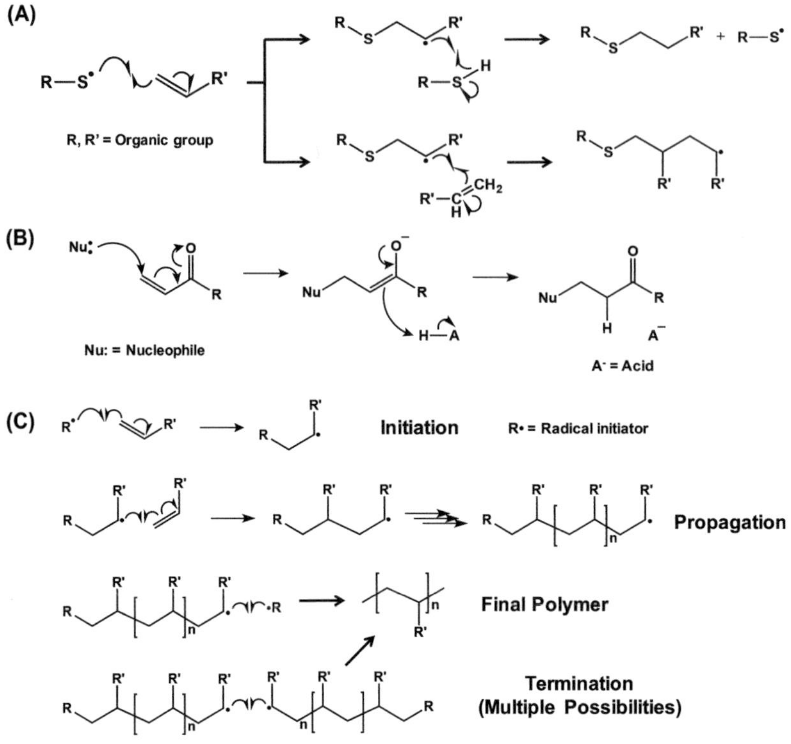

Fig. 10 Chemistries for covalent conjugation of peptide to polysaccharide backbone. (**a**) Thiol-ene reaction, (**b**) Michael-type addition, (**c**) radical polymerization

molecules in the RNA sample that are PCR inhibitors. Thus, NSC isolation described above requires customization. RT-qPCR for HA hydrogel systems have similar issues during RT-qPCR analysis with a much lower success rate (data not shown), and they need to be further optimized. The above-mentioned points need to be carefully considered when developing 3D NSC culture systems out of polysaccharide-based hydrogels.

The most ideal synthetic hydrogels are the ones installed with cell-adhesive ligands and, in addition, have physiologically relevant mechanical properties. The sugar polymer-based hydrogels, for that reason, offer a unique opportunity for researchers since the chemical side groups such as primary alcohols, carboxylic acids, and amines allow further modifications that can lead to the fine tuning of various physicochemical properties that are important for cell culture (i.e., cell adhesion, Young's modulus). As an example, thiol-ene, Michael-type addition, and radical addition chemistries (Fig. 10) are being utilized to covalently attach cell adhesive ligands (peptides) to the polymer backbone structures [37] while photo-initiator generated radicals and ionic cross-linking methodologies are utilized to tune the mechanical properties of hydrogels. Ultimately, synthetic (engineering) approaches require feedback from results gathered in the biology testing phase. Gene and protein expression levels will need to be precisely defined based on how the stem cells are being cultured (or differentiated) within the 3D microenvironment.

Acknowledgments

This work has been funded by Temple University and the GDFI Braincell Laboratory Co., Ltd., Seoul, South Korea.

References

1. Reynolds BA, Tetzlaff W, Weiss S (1992) A multipotent EGF-responsive striatal embryonic progenitor cell produces neurons and astrocytes. J Neurosci 12(11):4565–4574
2. Reynolds BA, Weiss S (1992) Generation of neurons and astrocytes from isolated cells of the adult mammalian central nervous system. Science 255(5052):1707–1710
3. Luskin MB (1993) Restricted proliferation and migration of postnatally generated neurons derived from the forebrain subventricular zone. Neuron 11(1):173–189
4. Lois C, Alvarez-Buylla A (1993) Proliferating subventricular zone cells in the adult mammalian forebrain can differentiate into neurons and glia. Proc Natl Acad Sci U S A 90 (5):2074–2077
5. Doetsch F, García-Verdugo JM, Alvarez-Buylla A (1997) Cellular composition and three-dimensional organization of the subventricular germinal zone in the adult mammalian brain. J Neurosci 17(13):5046–5061
6. Gage FH, Kempermann G, Palmer TD, Peterson DA, Ray J (1998) Multipotent progenitor cells in the adult dentate gyrus. J Neurobiol 36 (2):249–266
7. Gage FH (2000) Mammalian neural stem cells. Science 287(5457):1433–1438
8. Temple S (2001) The development of neural stem cells. Nature 414(6859):112–117
9. Doetsch F, Hen R (2005) Young and excitable: the function of new neurons in the adult mammalian brain. Curr Opin Neurobiol 15 (1):121–128
10. Carleton A, Petreanu LT, Lansford R, Alvarez-Buylla A, Lledo P-M (2003) Becoming a new neuron in the adult olfactory bulb. Nat Neurosci 6(5):507–518
11. Belluzzi O, Benedusi M, Ackman J, LoTurco JJ (2003) Electrophysiological differentiation of new neurons in the olfactory bulb. J Neurosci 23(32):10411–10418
12. van Praag H et al (2002) Functional neurogenesis in the adult hippocampus. Nature 415 (6875):1030–1034
13. Jessberger S, Kempermann G (2003) Adult-born hippocampal neurons mature into activity-dependent responsiveness. Eur J Neurosci 18(10):2707–2712
14. Ge S et al (2006) GABA regulates synaptic integration of newly generated neurons in the adult brain. Nature 439(7076):589–593
15. Addington C et al (2015) Enhancing neural stem cell response to SDF-1α gradients through hyaluronic acid-laminin hydrogels. Biomaterials 72:11–19
16. Stabenfeldt SE, García AJ, LaPlaca MC (2006) Thermoreversible laminin-functionalized hydrogel for neural tissue engineering. J Biomed Mater Res A 77(4):718–725
17. Jin K et al (2010) Transplantation of human neural precursor cells in Matrigel scaffolding improves outcome from focal cerebral ischemia after delayed postischemic treatment in rats. J Cereb Blood Flow Metab 30(3):534–544
18. Uemura M et al (2010) Matrigel supports survival and neuronal differentiation of grafted embryonic stem cell-derived neural precursor cells. J Neurosci Res 88(3):542–551
19. Banerjee A et al (2009) The influence of hydrogel modulus on the proliferation and differentiation of encapsulated neural stem cells. Biomaterials 30(27):4695–4699
20. Li X et al (2006) Culture of neural stem cells in calcium alginate beads. Biotechnol Prog 22 (6):1683–1689
21. Prang P et al (2006) The promotion of oriented axonal regrowth in the injured spinal cord by alginate-based anisotropic capillary hydrogels. Biomaterials 27(19):3560–3569

22. Ashton RS, Banerjee A, Punyani S, Schaffer DV, Kane RS (2007) Scaffolds based on degradable alginate hydrogels and poly (lactide-co-glycolide) microspheres for stem cell culture. Biomaterials 28(36):5518–5525
23. Wadowsky RM, Laus S, Libert T, States SJ, Ehrlich GD (1994) Inhibition of PCR-based assay for Bordetella pertussis by using calcium alginate fiber and aluminum shaft components of a nasopharyngeal swab. J Clin Microbiol 32 (4):1054–1057
24. White EJ, Venter M, Hiten NF, Burger JT (2008) Modified cetyltrimethylammonium bromide method improves robustness and versatility: the benchmark for plant RNA extraction. Biotechnol J 3(11):1424–1428
25. Wang L, Stegemann JP (2010) Extraction of high quality RNA from polysaccharide matrices using cetyltrimethylammonium bromide. Biomaterials 31(7):1612–1618
26. Köster N et al (2016) Single-step RNA extraction from different hydrogel-embedded mesenchymal stem cells for quantitative reverse transcription-polymerase chain reaction analysis. Tissue Eng Part C Methods 22 (6):552–560
27. Donato R et al (2007) Differential development of neuronal physiological responsiveness in two human neural stem cell lines. BMC Neuroscience 8:11
28. Burdick JA, Chung C, Jia X, Randolph MA, Langer R (2005) Controlled degradation and mechanical behavior of photopolymerized hyaluronic acid networks. Biomacromolecules 6 (1):386–391
29. Seidlits SK et al (2010) The effects of hyaluronic acid hydrogels with tunable mechanical properties on neural progenitor cell differentiation. Biomaterials 31(14):3930–3940
30. Williams CG, Malik AN, Kim TK, Manson PN, Elisseeff JH (2005) Variable cytocompatibility of six cell lines with photoinitiators used for polymerizing hydrogels and cell encapsulation. Biomaterials 26(11):1211–1218
31. Seidlits S et al (2010) The effects of hyaluronic acid hydrogels with tunable mechanical properties on neural progenitor cell differentiation. Biomaterials 31(14):3930–3940
32. Cao N, Chen X, Schreyer D (2012) Influence of calcium ions on cell survival and proliferation in the context of an alginate hydrogel. ISRN Chem Eng. doi:10.5402/2012/516461
33. Wilson IG (1997) Inhibition and facilitation of nucleic acid amplification. Appl Environ Microbiol 63(10):3741
34. Monteiro L et al (1997) Complex polysaccharides as PCR inhibitors in feces: Helicobacter pylori model. J Clin Microbiol 35(4):995–998
35. Demeke T, Adams R (1992) The effects of plant polysaccharides and buffer additives on PCR. Biotechniques 12(3):332–334
36. Westhrin M et al (2015) Osteogenic differentiation of human mesenchymal stem cells in mineralized alginate matrices. PLoS One 10 (3):e0120374
37. Sawicki L, Kloxin A (2014) Design of thiol-ene photoclick hydrogels using facile techniques for cell culture applications. Biomat Sci 2(11):1612–1626

Neuromethods (2017) 128: 205–211
DOI 10.1007/7657_2017_2

Published online: 30 April 2017

Measuring miRNA Mediated Translational Regulation with Live Cell Imaging

Min Jeong Kye

Abstract

Numerous microRNAs are detected in synaptic areas such as axons and dendrites. As accurate regulation of local protein synthesis can be crucial for the function of neurons, miRNAs can play very important roles for mRNA translation in the rather isolated cellular spaces such as synaptic area. However, due to the technical limitation, it is very difficult to measure efficiency of protein synthesis in the synaptic area with biochemical methods. Therefore, visualizing translation and measuring protein levels at the synaptic sites by imaging techniques can be a good alternative. Fluorescence recovery after photobleaching (FRAP) has been widely used to measure local protein synthesis rate in axons. This technique allows us to measure speed and efficiency of translation of certain mRNAs. We modified this method to measure how miRNAs influence on local synthesis of proteins in neurons.

Keywords: MicroRNA, Local translation, Axon, FRAP (fluorescence recovery after photobleaching), Live cell imaging

1 Introduction

Numerous microRNAs are detected in synaptic areas such as axons and dendrites [1–3]. Due to the highly sophisticated morphology and quick signaling pathways at the synapse, neurons synthesize proteins at the local site of subcellular compartments such as axons and dendrites [4]. However, because of their highly polarized morphology, it has been challenging to measure the efficiency of local protein synthesis in neuron. Various methods have been intensively developed to measure the efficiency of protein synthesis [5–9]. First, in brief, newly synthesized proteins can be labeled by amino acid azidohomoalanine (AHA) in hippocampal neurons. This showed that dendritic protein synthesis was elevated after brain derived neurotrophic factor (BDNF) treatment. This method allows us to visualize newly synthesized proteins in subcellular compartment of neurons in situ. Moreover, it confirms that neurotrophic factor, BDNF, increases local protein synthesis [5]. This method has been widely used to measure in vivo protein synthesis in various systems including *Drosophila* models of neurodegenerative disorder, Charcot–Marie–Tooth neuropathy [10]. Second widely used method is fluorescence recovery after photobleaching (FRAP) [8]. In this method, coding region of reporter protein is fused to 3′ of untranslated region of target mRNA. The reporter proteins are photo-destructible fluorescent proteins, and they are usually myristoylated to reduce/minimize diffusion. First of all, if the target mRNA is locally translated in axons or dendrites, the signal of the reporter protein can be detected in subcellular compartments. After that, signals can be photobleached by strong laser, and recovery of the reporter signal, which means newly synthesized proteins, can be measured by microscope [8]. This method has been widely modified and used in various model systems. This can measure the efficiency of axonal local protein synthesis in different conditions, including genetic background or environmental cues [11–13]. Moreover, reporter protein fused to the target protein can be used to measure local protein level in axons and dendrites [14, 15]. Using this method, the role of miRNAs for local translation can be measured [16].

Here, we will describe the detailed protocol for measuring miRNA related protein synthesis in axons using FRAP.

2 Materials

2.1 Primary Neuron Culture

- Embryonic mouse (E18 for hippocampal neurons)
- Poly-D-lysine (Sigma-Aldrich)
- Trypsin (Worthington)
- DNase I (Worthington)
- Neurobasal media (Thermo Scientific)
- Minimum Essential Media (Thermo Scientific)
- Glucose (Sigma-Aldrich)
- Fetal Calf Serum (Biochrome)
- B27 (Thermo Scientific)
- Pen/Strep (Thermo Scientific)
- PBS (Thermo Scientific)
- Tweezers
- Scissors
- Dissecting microscope

2.2 Vectors

- Reporter (myristoylated Green Fluorescent Protein) vectors fused with 3′ untranslated region of target mRNA

2.3 Microscope for Live Cell Imaging System

- Zeiss LSM5 confocal microscopy with 37 °C platform
- Imaging culture plate (35-mm bottom glass dishes, MatTek)
- Hibernate E Low Fluorescence medium (Brain Bits LLC)

2.4 Nucleofection

- Amaxa nucleofector (Lonza)
- Amaxa nucleofector kit for primary neural cells (Lonza)

2.5 Chemicals

- Anisomycin (Protein synthesis inhibitor, Sigma-Aldrich)
- Monastrol (KIF5 inhibitor, Sigma-Aldrich)

3 Methods

3.1 Preparing Primary Neurons for Imaging

(a) Hippocampi were dissected from E18 mouse embryos.

(b) Neurons were dissociated with trypsin (0.01 % for 5 min at RT), triturated with DNase I (0.1 %) in plating media (MEM, 0.6 % glucose, 5 % FCS, and 1× Pen/Strep).

(c) 1 μg of plasmid DNA (myristoylated GFP fused with 3′ UTR of target mRNAs) were transfected to 500,000 neurons using AMAXA nucleofection.

(d) 100,000 neurons were plated on poly-D-lysine coated 35 mm glass bottom imaging dishes. Neurons grow in humidified 37 °C incubator.

(e) Plating media was replaced to maintenance media (neurobasal media, B27, and 1× Pen/Strep) in 4 h after plating.

(f) Good images can be taken between 48 and 72 h after plating.

3.2 Performing Fluorescence Recovery After Photobleaching

3.2.1 Testing Target mRNAs for Fluorescence Recovery After Photobleaching

First of all, to be locally translated, target mRNA needs to be trafficked to subcellular compartments of neurons such as axons and dendrites. Therefore, it is important to check whether signal of reporter can be detected in axons. GFP signal can be observed in axon after 48 h of transfection (Fig. 1). As GFP molecules are myristoylated, diffusion of GFP protein is at minimum level. Thus, we assume that GFP proteins in axonal compartment are locally synthesized.

3.2.2 Optimizing Fluorescence Recovery After Photobleaching Conditions

(a) Neurons were washed and switched to pre-equilibrated Hibernate E Low Fluorescence medium (Brain Bits LLC).

(b) FRAP was performed using a Zeiss LSM5 upright confocal - microscope fitted with a heating stage and 63× water-immersion objective (numeric aperture = 0.9). A rainbow spectrum display was employed to adjust the gain and offset to ensure that the highest pixel intensity is just saturating (Fig. 2).

(c) The region of interest containing axons/growth cone was bleached using 100 % laser power for 120 iterations (<2 min), followed by regular imaging using 15 % laser power at speed of

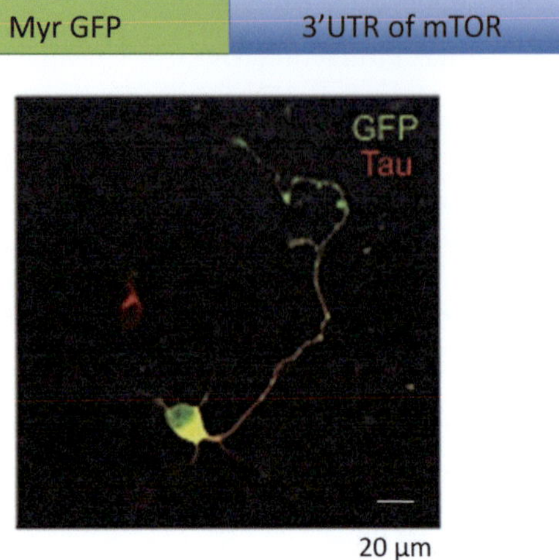

Fig. 1 cDNA of myrGFP is fused with 3′ UTR of target genes. GFP signal can be found in the axon of 3 days cultured neurons. Scale bar 20 μm

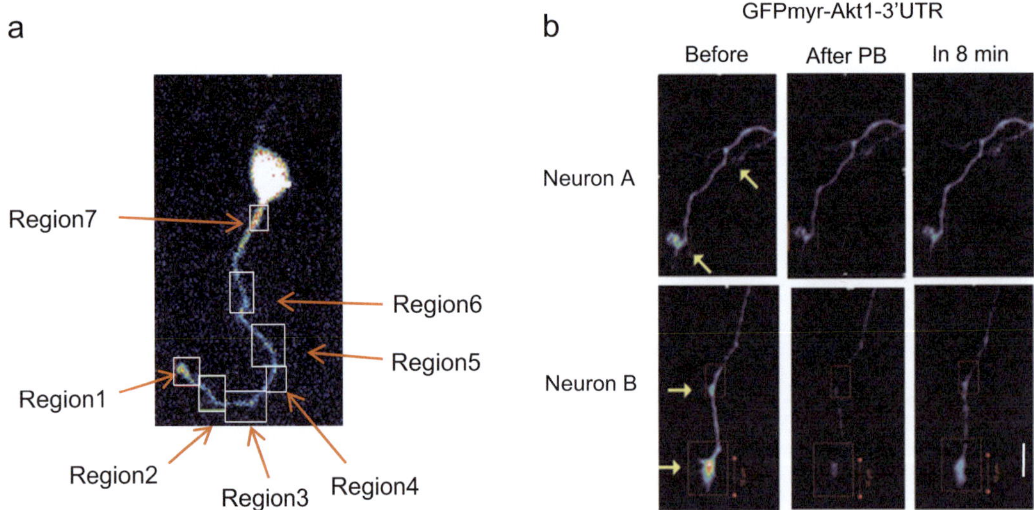

Fig. 2 (**a**) One example of fluorescence recovery after photobleaching (FRAP) experiment of myrGFP fused with 3′ UTR of *Akt1*. We defined several regions in growing axon from growth cones to cell body (soma) and photobleached every regions of axonal compartment of neurons. The recovery rates of signals from individual regions are measured to check how much of the signals are recovered after photobleaching. (**b**) Strong signal of translation was observed in growth cone and branching point of growing axons suggesting that these are translational hot spots. Scale bar 10 μm

7.86 s per scan with a 52.3-s interval over a 10-min course (Fig. 2). Data were collected by Zeiss program, Zen, and analyzed manually.

(d) As a proof of principle, to test FRAP signals are obtained from mRNA translation, neurons were pretreated with 100 μM of anisomycin or DMSO for 30 min before performing FRAP (Fig. 3).

(e) To test possible trafficking of GFP molecules from soma (cell body) to axon, the cultures were incubated with 50 μM monastrol, which blocks anterograde axonal trafficking of granules and vesicles by inhibiting KIF5 for 30 min prior imaging and performed FRAP (Fig. 3).

(f) Fluorescence intensity was measured using ImageJ and statistical analyses were performed in Prism 5 using two-way ANOVA to compare significant signal recovery over the course of treatment and among different treatments.

3.2.3 Testing Whether miRNA Regulates Local mRNA Translation

(a) To test whether certain miRNA regulates translation of your target mRNA, 1 ug of plasmid DNA and 50 pmole of control or miRNA inhibitor (locked nucleic acids, LNAs) were transfected together using AMAXA nucleofection.

(b) As it is described above, within 48–72 h after nucleofection, FRAP can be performed to measure active axonal local translation.

(c) If miRNA regulates mRNA translation, modifying miRNA expression can change FRAP recovery rate (Fig. 4).

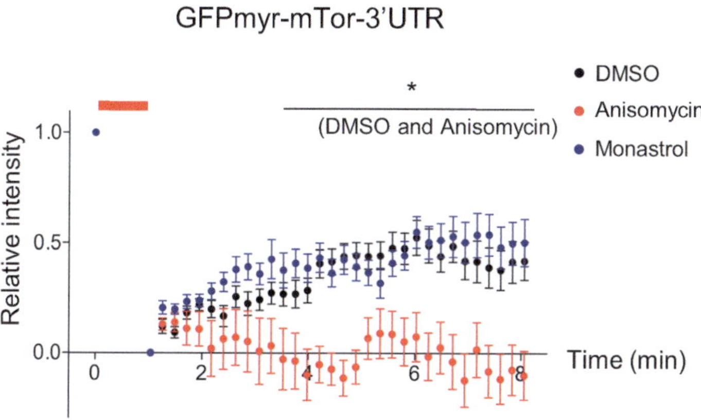

Fig. 3 Quantification of GFP intensity of dGFPmyr-*mTor*-3′ UTR in the distal axons and growth cones before and after photobleaching (PB). Anisomycin (50 μM, *red*) blocked recovery of GFP signal after PB. Monastrol (anterograde axonal trafficking inhibitor, 50 μM, *blue*) did not exhibit any significant effects on recovery of GFP signal compared to vehicle treated neurons (DMSO, *black*). For each time point, the data represent a relative ratio of prebleach intensity (mean ± S.E.M). DMSO: $n = 33$ regions from 11 neurons, anisomycin: $n = 17$ regions from five neurons, and monastrol: $n = 19$ regions from six neurons. Statistical significance was determined by two-way ANOVA followed by Bonferroni's post hoc test. Anisomycin: $^*p < 0.05$ from 3.99 to 8.06 min (anisomycin data are published in Kye et al. [16])

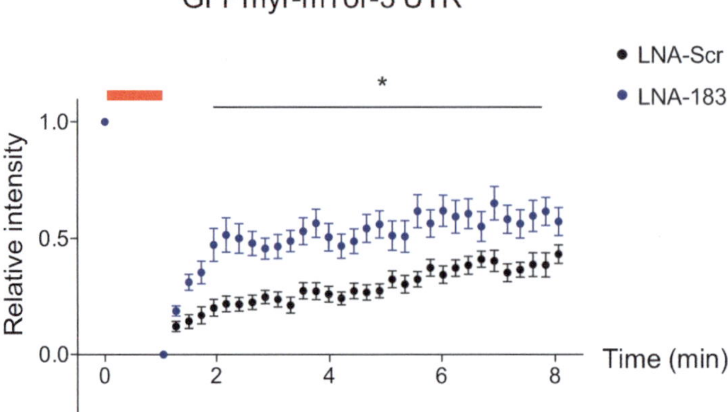

Fig. 4 *miR-183* inhibited neurons showed significantly increased recovery of GFP signal after PB. For *mTor*-3′ UTR, control (LNA-Scr, *black*): $n = 23$ from seven neurons and miR-183 inhibition (LNA-183, *red*): $n = 22$ from five neurons. $^{*}p < 0.05$ from 1.95 to 7.84 min. *Red bars* represent times of photobleaching

4 Conclusions

Due to the morphological complexity of neurons and difficulties to obtain pure materials from synaptic area, it has been challenging to have clear answer about biochemical reactions at the synapse. Many synaptic microRNAs have been listed [2, 17, 18]. However, it has been difficult to characterize how/which/when miRNAs work in the synaptic area. Various approaches have been made to visualize local protein synthesis. Each method has its advantages and disadvantages. For example, enzymatic method using amino acid substitute can be clearly quantitative; however, it cannot be used for specific mRNA. While FRAP can be more gene specific, experimental system is extremely artificial. Here we summarized how we perform FRAP to check how miRNA can influence on mRNA translation in axonal compartment of growing neurons. This is still not the optimal method, as neuronal compartment specific miRNA knockdown method is unavailable at the moment. However, this can suggest that miRNAs regulate local translation.

Acknowledgement

This work is supported by Deutsche Forschungsgemeinschaft (KY-92/1), University of Cologne (Cologne Fortune), and Cure SMA.

References

1. Schratt GM, Tuebing F, Nigh EA, Kane CG, Sabatini ME, Kiebler M, Greenberg ME (2006) A brain-specific microRNA regulates dendritic spine development. Nature 439(7074):283–289. doi:10.1038/nature04367

2. Kye MJ, Liu T, Levy SF, Xu NL, Groves BB, Bonneau R, Lao K, Kosik KS (2007) Somatodendritic microRNAs identified by laser capture and multiplex RT-PCR. RNA 13(8):1224–1234. doi:10.1261/rna.480407
3. Natera-Naranjo O, Aschrafi A, Gioio AE, Kaplan BB (2010) Identification and quantitative analyses of microRNAs located in the distal axons of sympathetic neurons. RNA 16(8):1516–1529. doi:10.1261/rna.1833310
4. Holt CE, Schuman EM (2013) The central dogma decentralized: new perspectives on RNA function and local translation in neurons. Neuron 80 (3):648–657. doi:10.1016/j.neuron.2013.10.036
5. Dieterich DC, Hodas JJ, Gouzer G, Shadrin IY, Ngo JT, Triller A, Tirrell DA, Schuman EM (2010) In situ visualization and dynamics of newly synthesized proteins in rat hippocampal neurons. Nat Neurosci 13(7):897–905. doi:10.1038/nn.2580
6. Taylor AM, Dieterich DC, Ito HT, Kim SA, Schuman EM (2010) Microfluidic local perfusion chambers for the visualization and manipulation of synapses. Neuron 66(1):57–68. doi:10.1016/j.neuron.2010.03.022
7. Wu B, Eliscovich C, Yoon YJ, Singer RH (2016) Translation dynamics of single mRNAs in live cells and neurons. Science 352(6292):1430–1435. doi:10.1126/science.aaf1084
8. Vuppalanchi D, Coleman J, Yoo S, Merianda TT, Yadhati AG, Hossain J, Blesch A, Willis DE, Twiss JL (2010) Conserved 3′-untranslated region sequences direct subcellular localization of chaperone protein mRNAs in neurons. J Biol Chem 285(23):18025–18038. doi:10.1074/jbc.M109.061333
9. Yoon YJ, Wu B, Buxbaum AR, Das S, Tsai A, English BP, Grimm JB, Lavis LD, Singer RH (2016) Glutamate-induced RNA localization and translation in neurons. Proc Natl Acad Sci U S A 113(44):E6877–E6886. doi:10.1073/pnas.1614267113
10. Niehues S, Bussmann J, Steffes G, Erdmann I, Kohrer C, Sun L, Wagner M, Schafer K, Wang G, Koerdt SN, Stum M, Jaiswal S, RajBhandary UL, Thomas U, Aberle H, Burgess RW, Yang XL, Dieterich D, Storkebaum E (2015) Impaired protein translation in Drosophila models for Charcot-Marie-Tooth neuropathy caused by mutant tRNA synthetases. Nat Commun 6:7520. doi:10.1038/ncomms8520
11. Akten B, Kye MJ, Hao le T, Wertz MH, Singh S, Nie D, Huang J, Merianda TT, Twiss JL, Beattie CE, Steen JA, Sahin M (2011) Interaction of survival of motor neuron (SMN) and HuD proteins with mRNA cpg15 rescues motor neuron axonal deficits. Proc Natl Acad Sci U S A 108(25):10337–10342. 1104928108 [pii]. doi:10.1073/pnas.1104928108
12. Rathod R, Havlicek S, Frank N, Blum R, Sendtner M (2012) Laminin induced local axonal translation of beta-actin mRNA is impaired in SMN-deficient motoneurons. Histochem Cell Biol 138(5):737–748. doi:10.1007/s00418-012-0989-1
13. Nie D, Di Nardo A, Han JM, Baharanyi H, Kramvis I, Huynh T, Dabora S, Codeluppi S, Pandolfi PP, Pasquale EB, Sahin M (2010) Tsc2-Rheb signaling regulates EphA-mediated axon guidance. Nat Neurosci 13(2):163–172. doi:10.1038/nn.2477
14. Spinelli KJ, Taylor JK, Osterberg VR, Churchill MJ, Pollock E, Moore C, Meshul CK, Unni VK (2014) Presynaptic alpha-synuclein aggregation in a mouse model of Parkinson's disease. J Neurosci 34(6):2037–2050. doi:10.1523/JNEUROSCI.2581-13.2014
15. Banerjee S, Neveu P, Kosik KS (2009) A coordinated local translational control point at the synapse involving relief from silencing and MOV10 degradation. Neuron 64(6):871–884. doi:10.1016/j.neuron.2009.11.023
16. Kye MJ, Niederst ED, Wertz MH, Goncalves Ido C, Akten B, Dover KZ, Peters M, Riessland M, Neveu P, Wirth B, Kosik KS, Sardi SP, Monani UR, Passini MA, Sahin M (2014) SMN regulates axonal local translation via miR-183/mTOR pathway. Hum Mol Genet 23(23):6318–6331. doi:10.1093/hmg/ddu350
17. Risbud RM, Porter BE (2013) Changes in microRNA expression in the whole hippocampus and hippocampal synaptoneurosome fraction following pilocarpine induced status epilepticus. PLoS One 8(1):e53464. doi:10.1371/journal.pone.0053464
18. Boese AS, Saba R, Campbell K, Majer A, Medina S, Burton L, Booth TF, Chong P, Westmacott G, Dutta SM, Saba JA, Booth SA (2016) MicroRNA abundance is altered in synaptoneurosomes during prion disease. Mol Cell Neurosci 71:13–24. doi:10.1016/j.mcn.2015.12.001

Neuromethods (2017) 128: 213
DOI 10.1007/7657_2016_10

Published online: 18 October 2016

Erratum to: Experimental Methods for Functional Studies of microRNAs in Animal Models of Psychiatric Disorders

Vladimir Jovasevic and Jelena Radulovic

Erratum to: Neuromethods
DOI 10.1007/7657_2016_9

The publisher regrets that an incorrect version of Fig. 4 was included in the chapter. The correct figure is shown below.

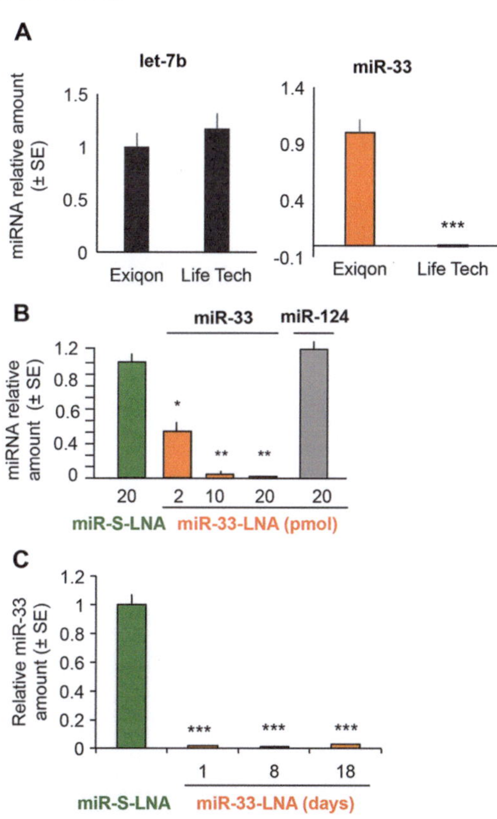

The updated original online version for this book can be found at DOI 10.1007/7657_2016_9

Neuromethods (2017) 128: 215–216
DOI 10.1007/978-1-4939-7175-6

Index

N

P

Q

R

S

T

MIX
Papier aus verantwortungsvollen Quellen
Paper from responsible sources
FSC® C105338

If you have any concerns about our products,
you can contact us on
ProductSafety@springernature.com

In case Publisher is established outside the EU,
the EU authorized representative is:
Springer Nature Customer Service Center GmbH
Europaplatz 3, 69115 Heidelberg, Germany

Printed by Libri Plureos GmbH
in Hamburg, Germany